D0773373

Advance Praise

"Collin McLoughlin has written a must-read primer for anyone considering a Lean turnaround and a complete guide to implement the best productivity tools available for any industry… Read this book—and learn from one of the best books I've ever read."

Luis Socconini
President, Lean Six Sigma Institute
CEO/Master Black Belt
Lean Six Sigma Institute
www.leansixsigmainstitute.org

"In writing *True Kaizen*, Collin McLoughlin and Toshihiko Miura have created an insight into Kaizen that cleverly intertwines its practical application with its underlying philosophy. A great example of this was the explanation of tackling the 'Elephant in the Room', a metaphor that most of us are familiar with, but by providing insight into the way that Taiichi Ohno managed these types of scenarios. McLoughlin and Miura are true practitioners of Lean thinking and I can recommend this book as a 'must-read' for anyone wishing to embed a Kaizen culture into their organization. You don't often get the opportunity to peek inside of the minds of two great business advisors, and this is one opportunity that you must take."

Philip Holt
Head of Operational Excellence
Accounting Operations
Royal Philips
www.philips.com

True Kaizen is a guide into the mindset of Kaizen as originally envisioned by Taiichi Ohno. The book focuses on three critical elements of leadership and management: (1) addressing and improving the perspective (mindset) of management, (2) exploring and improving the workplace environment, and (3) observing and eliminating wasteful work processes. While there have been many publications focused on the third element, far too little has been written about the mindset of leaders and the impact this can have on the work environment. The principles throughout the book are illustrated by numerous examples from companies in many industries in Japan and other countries. I am thankful that Collin McLoughlin and Toshihiko Miura have transcribed the teachings of Mr. Ohno and Mr. Yamada into clear and logical lessons that leaders everywhere will be able to use to improve the morale and performance of their workplace.

Lynn D. Martin, MD, MBA
Medical Director, Continuous Performance Improvement
Interim Director, Bellevue Surgery Center
Seattle Children's
Professor of Anesthesiology and Pediatrics (Adj.)
University of Washington School of Medicine
http://www.seattlechildrens.org/

True Kaizen

Management's Role in Improving Work Climate and Culture

True Kaizen

Management's Role in Improving Work Climate and Culture

Collin McLoughlin
Toshihiko Miura

CRC Press is an imprint of the
Taylor & Francis Group, an **informa** business
A PRODUCTIVITY PRESS BOOK

CRC Press
Taylor & Francis Group
6000 Broken Sound Parkway NW, Suite 300
Boca Raton, FL 33487-2742

Printed on acid-free paper

International Standard Book Number-13: 978-1-138-74542-1 (Hardback)
International Standard Book Number-13: 978-1-315-18037-3 (eBook)

Visit the Taylor & Francis Web site at
http://www.taylorandfrancis.com

and the CRC Press Web site at
http://www.crcpress.com

Contents

List of Figures

Foreword

It seems like such common sense, the idea that organizations are more adaptive and more successful when they engage everybody in improvement. So why is that approach so uncommon?

When I started my career at General Motors over 20 years ago, our engine plant performed very poorly compared to the benchmark Toyota plant. Quality was poor, productivity was almost exactly half that of Toyota, and workplace morale was extremely low. Perhaps I should list "low morale" first there, since that was undoubtedly one of the causes of low quality and high cost. But, the problem was not the workers; it was our management system.

Our GM plant built a very similar product with the same machines (purchased from the same vendors). The machines were physically laid out like Toyota in an attempt to copy their success. The frontline workers had the same educational background as our competitors. The only meaningful variable was a difference in leadership styles. One fault with the old approach was the arrogant notion that managers always knew best and the workers should just keep quiet. "Management tells us to check our brains at the door" was one of the common complaints in that environment.

When did our plant begin to turn around? After a number of major quality incidents, we got a new plant manager who had been one of the first GM managers to be assigned to work with Toyota at the famed NUMMI joint venture plant in California. At NUMMI, he had learned a new way of managing and that mindset was brought into our plant. The culture started to change as we respected and engaged the workforce. Within a few short years, the plant quickly moved up from being the "worst of the worst" to the top quartile of performance. The plant didn't replace the workers or the equipment; it changed the culture.

A culture of continuous improvement, a team or an organization where "kaizen" has become the common mindset, is such a joyful, beautiful thing to see and experience—whether it's a manufacturing shop floor, a nonprofit organization, or a hospital. It's sad to hear anybody complain about their ideas not being listened to, but it's particularly heartbreaking to hear the same lament from nurses, doctors, and other healthcare professionals.

If it makes good sense that we should respect and engage people, what then are the barriers to a culture of improvement? We are often our own worst enemy when it comes to self-defeating, negative thoughts—we don't have time, our employees won't have good ideas, they won't participate, it won't make a difference. The good news is that, with sustained leadership and a lot of diligence, we can shift the discussion from "here's why we cannot" to "how might we?"

"How might we?" is a powerful phrase and it is part of the culture at Franciscan Health in Indianapolis, where my *Healthcare Kaizen* coauthor Joe Swartz has led their improvement efforts and culture change for over a decade. Leaders and staff there don't just react to problems or waste; they challenge themselves to get better. They ask, "How might we improve?" How might we achieve the goals that matter to our customers, our colleagues, and our organizations? The sense of challenge and the positivity that's implied in trying to improve is a theme in this book, as well.

I'm pleased and honored to write the Foreword for the book *True Kaizen*. I believe this book will be a helpful addition to the literature on Toyota, Lean, and continuous improvement. I have known Collin McLoughlin for a decade and he's very committed to furthering his own knowledge and helping others discover that joy of improvement. His coauthor Toshihiko Miura is well positioned to help the reader translate the spirit and practice of Kaizen from Japan to your location and industry.

Kaizen is, thankfully, not a uniquely Japanese mindset or practice. I've learned, in just two visits to Japan, that it's not the default culture in every Japanese business. This book shares the philosophy and practical examples of how organizations in any country or any industry can create a culture of improvement.

I've seen "True Kaizen" in healthcare organizations all around the world. In our increasingly connected world, inspiring stories of Kaizen get shared on YouTube and social media, which helps inspire others to rethink their own workplaces.

In this book, you'll learn that Kaizen success does not depend on the format of a card or the exact labels on a huddle board; it's about leadership. *True Kaizen* will help you understand the mindsets, the philosophy and the leadership behaviors that will create and sustain a culture of improvement.

These mindsets aren't rocket science, as they say, but they will challenge the way things are normally done in many organizations. In my experience and practice, Kaizen involves a shift from *knowing* answers in a definitive way to a process of *finding* and *testing* answers in practice. As they say at Franciscan Health, "We don't know if it's a good idea until we see if it works." Many leaders are also stuck in a mindset that says people will only improve if they set targets or provide incentives. In a culture of continuous improvement, leaders learn, instead, to tap into the intrinsic motivation of those who are doing the work, in the name of improving their own work, making things better for patients or customers, and helping their organization. Needless to say, trust is an underlying requirement, as you'll read about in this book.

As the authors say, it's better for 100 people to take a small step forward, than it is for a single person to take 100 steps. Reading this book will inspire and enable you to take your first step forward, while showing you how to get others to also take their first step toward a culture of improvement.

This book contains thoughtful insights that will allow you to reframe improvement from being a burden to a generator of smiles, pride, and better business results. Kaizen will shift from being some new program to being a way of thinking and a way of life. You'll more than likely start practicing Kaizen at home out of your own intrinsic motivation to make your life better. Before long, you won't be able to imagine going back to the old way of leading and managing.

Mark Graban
Author of Lean Hospitals
Coauthor of Healthcare Kaizen

Preface

THE MANAGEMENT EQUATION

What's the magic formula to achieving effective management? Walk down any aisle in the business section of a bookstore and you will find numerous books claiming to have found the solution to all management's problems. These books typically start with famous steps, such as Franklin Covey's *The 7 Habits of Highly Effective People* or 10 Steps-to-Something. And who wouldn't want a quick fix to a problem they don't see an easy solution for?

However, the truth of the matter is that the most successful management and leadership principles cannot be broken down so easily into a short checklist. There are many paths that lead to management success, and if you were to write down all of the nuances to every management style, you could cover the globe in pages and still not be done writing.

Given the fact that there is no magic formula written down, one might think, "well, let's take a look at successful companies and see what they did," and while this is one way to learn what people and organizations have accomplished, this method is not the magic answer either. Even looking back over the past few decades, you will see that hundreds of companies (even those very famous) rise and eventually fall within a relatively short span of years.

The reasons for the rise and fall of each company differ, due to issues such as changing political climates, changes or leaps in technology, legal pressures due to company size changes, and changes in the market competition and supply chain. However, one thing is constant across all companies, regardless of size or location: at one point in the organization's history, the management team had to struggle and learn how to balance all these factors, while also looking forward and anticipating the future.

Making this balancing act even more challenging is the fact that all organizations are made up of hundreds of small parts, much like a complex machine. These small parts include technology, raw materials, discrete operations, human resources, and multiple communication channels. Each one of these elements must deal with a multitude of challenges every day, from the leadership style of the executive team, to organizational design changes, local and international culture interactions, and even the understanding and adoption (or buy-in) of core organizational values by its employees. All of these factors combined create a very complicated puzzle that must be solved in order for the organization to survive.

When taking all of this complexity into account, it is easy to understand how a single management book cannot be the answer to everyone's management challenges. An observer, even a person writing a book on management observations, cannot cover all the nuances of each person within the organization in order to come to the conclusion of success.

To gain even a glimpse of what it takes to manage an organization to success, we need to take a step back and truly focus on the human element of organizations. As many people have pointed out in numerous articles and journals over the past decades, we have to recognize that organizations need to fulfill human needs and wants in order to experience success. The greatest organizations, those that have withstood the test of time, have focused on the people who make up the organization, their success, and what it takes to fulfill the lives of those employees.

PRODUCTIVITY

The idea of productivity is, in fact, a human measurement and needs to be considered as such. We cannot think about productivity without the human element.

Let's look at the human factor when considering organizational productivity, using a researcher named Walter Davis. Davis* highlighted the human factor with an excellent model, showing the importance of the human element to productivity (see Figure P.1).

Davis stated that knowledge multiplied by skill equals ability, that attitude multiplied by the situation equals motivation, and that ability multiplied by motivation equals potential human performance. Therefore,

* Organizational Behavior: Human Behavior at Work—Walter Davis, 9th Edition, McGraw-Hill, 1993.

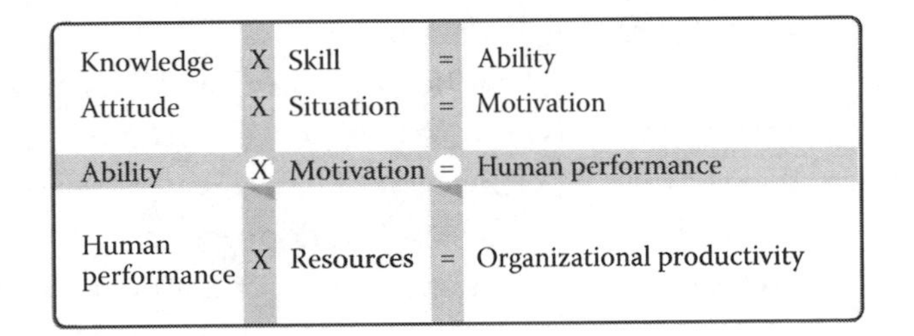

FIGURE P.1
Organizational performance equation.

potential human performance multiplied by resources available equals the organizational performance potential, or productivity.

While the point of this model seeks to explain organizational potential in terms of the human element, what it more importantly does is provide a perspective on how we define and measure productivity. That is why this formula has endured over time.

As our workforce becomes more knowledgeable, skillful, and perceptive of their needs and wants as employees, the ability to reach the true potential of an organization becomes more and more difficult.

As managers, we must look at each individual element of an equation like this in order to fully understand how we can achieve an answer. We must begin to answer more focused questions, such as the following:

1. How productive is the existing work climate and culture?
2. How do employees, as individuals, navigate the existing work climate? (How do they deal with day-to-day issues with each other?)
3. Where and how are individuals and their work processes assessed?
4. What obstacles do employees face every day, and are they empowered to fix these obstacles?
5. What role does leadership play at each level of the organization? (Looking at the organization in layers of management.)

To address these challenges, this book focuses on three main aspects of leadership and management.

1. Addressing and improving the perspective of management
   The ideas presented in this book are not limited to a certain industry or field of work but can be applied in any setting because they speak to a universal human element.

2. Exploring and improving work climate

   Organizations are social entities, operating within their own controlled environment. This book will explore the factors that contribute to, and encourage, a positive work climate.
3. Observing and eliminating wasteful work processes

   Observing wasteful activities and work processes requires a refined perspective. Through the case studies presented in this book, we will illustrate the how and why, in order to help you refine your expertise. This will also lead to the joy and benefits that come from a truly engaged workforce.

Although most of the examples in this book are set in manufacturing environments, these same concepts are being applied in hospitals, offices, and retail operations throughout the world. Our experience has shown that these practices transcend nations, cultures, and workplaces. We have personally seen them adopted in such diverse locations as shipping facilities in Germany, meat processing plants in New Zealand, and bakeries in China.

We know you will find value and relevance in the examples here for your own work. It is our hope that you begin to shift the way you look at work and question the why and how of things being done to achieve management and leadership success.

Authors

Collin McLoughlin has championed total organizational change across 12 different industries. His access to world-class Lean organizations and his real-world experience with True Kaizen have allowed him to serve as the chief architect of organizational change in some of the world's leading organizations. Collin is recognized for coaching and advising influential businesses and leaders around the world, including Carl Zeiss, Seattle Children's Hospital, Mars, Siemens, Heinz, and Dole, just to name a few.

Collin has led well over 50 Kaizen Study Missions to Japan to benchmark and to mentor organizational leadership. This has given him the insight to author more than 600 different books, training materials, workshop packages, and subject-based videos to help organizations succeed at creating sustainable improvement systems. He directs reformation of businesses around the globe and serves as president of Enna.com and as chief advisor to key organizations.

Toshihiko Miura has helped organizations around the world achieve Kaizen transformations in many industries, from manufacturing to healthcare and beyond. His Kaizen training is especially unique and invaluable as he first focuses on developing morale and the engagement level of each worker. After he confirms there is a true commitment toward Kaizen from every individual within an organization, he then guides companies through meaningful transformation.

Miura has learned the true meaning of Kaizen from Sensei Hitoshi Yamada, who is a direct student of the inventor of The Toyota Production System, Taiichi Ohno. Miura is the director of Yamada's Personal Education Center (PEC), which works to spread knowledge of the Kaizen philosophy across Japan, and he has been a contributor to many of their publications. His mission to spread this vital knowledge extends to all the organizations he works with, whether they are in Japan, Canada, Australia, or anywhere else in the world.

1

Why Leadership Matters

Consider some of the industrial giants that have fallen in the last 20 years. They had resources, their employees' varied intellectual talents, and very prominent positions in the corporate world. So why were they unable to successfully navigate through the conditions that faced them and remain successful?

To understand their success, and eventual failure, you must be able to first see how business strategy, implementation, and culture all have their roots in the corporate mindset and values of an organization. Leadership is the key to generating a corporate mindset and therefore is also the key to performance.

For instance, when a sports team does not succeed on the grass, ice, or court, the media and fans look to the coach, as he is the leader. A famous basketball coach, Jon Wooden, stated that he was never worried about his opposition because it was what his own team did with the skills they had which determined their success or failure, not what the other team did or did not have.

Think about that statement in terms of your own organization. Do you focus on the talent and skills of the people who work there, or is there a preoccupation with looking at, and comparing with, your competition? Too often people get caught up in comparisons between themselves and others, instead of focusing on what they are good at. This is amplified in groups and the leadership of organizations, especially in very competitive environments.

Most organizations have access to the same general resources (people, materials, machines, and information systems). The competitive difference and strength of an organization lies in how these resources are put together, the mechanisms that people create based on these resources. This integration of methods, as well as focusing an organization's human resources to attain specific goals, is what determines success or eventual failure.

While the abilities to anticipate, react, extend, and (dare we say) bluff are all necessary skillsets to win in both sports and business, it is how well these skills are implemented and performed that determine success. Even more telling than skills, however, is the ability for people to make decisions without needing to escalate them up a layer of command and still be aligned with the organization or team's objectives.

The next time you watch a sporting event, carefully watch both the player's decisions on the field/rink/court and how the coach interacts and adjusts the overall game plan.

The coach will guide, reinforce, and make higher-level decisions on adjustments because he or she can see the larger picture (the whole game). Meanwhile, the players will be making split-second decisions at critical junctures to help advance the team toward the goal, while staying aligned to the newest direction from the coach. This is the power of decision making at each level of the group and highlights the importance of truly observing a situation before making any changes or adjustments to the current strategy. If the team is winning, then there is no need to change. However, if the team is struggling against their opponent, the coach must observe the challenges, make decisions to restructure and realign players, and then continue to observe to determine if their changes improve the overall team performance.

This need for observation, not assumption, is what we have learned through years of experience in observing and participating in the changes of a multitude of companies. Assumptions are based on personal experiences, not facts, and those assumptions are often the reason for poor decision making. Decisions should be made based on factual observation and information in order to have the greatest impact on the issue at hand.

Business leaders need to look at the field of play in much the same way a coach does. They must know their team, understand and comprehend the larger objective, and then make changes and adjustments to help guide the people in the organization toward the greater overall goal. They need to look beyond the day-to-day activities being performed and really look at their people as players who are constantly gaining new skills, new experiences, new talent, and even some aptitude that cannot be explained. In this way, business leaders will be able to strike that delicate balance of leadership and management.

MAINTAINING A HOLISTIC PERSPECTIVE

In order to provide effective leadership it is important to be able to diagnose, adapt, and communicate within the workplace.

Diagnosing

The first competency, diagnosing (or analyzing), means being able to understand the dynamics of a situation and see what needs to change and adapt in order to achieve success.

We have the privilege of spending weeks in Japan each year, guiding numerous leadership tours on which we teach managers how to assess an operation quickly in order to determine cost of goods sold as well as profit margins. This is a good basic starting point from an outside observer to understand how a business is operating. During these tours, we teach a method based on an article by Eugene Goodson (2002), called "How to Read a Plant Fast."*

"How to Read a Plant Fast" categorizes types of observations into 11 key categories, which can be used to determine an organization's vitality from culture to overall performance. These metrics consider such factors as cost, quality, safety, customer service, and productivity, all while not even looking at one income or balance sheet. This is important to understand, since these observations come from a different perspective than most people are used to taking. This adjusted perspective—that it is not about the numbers, but the more organic nature of the people working together—will often be the final mark as to how well a company performs.

This method is invaluable to gaining a new understanding of business and illustrates to the leaders on our trip why they need to know more than just their own business metrics to be successful and how they can acquire this knowledge.

Adapting

The second competency, adapting, is being able to apply principles in different contexts and business situations. It is important to understand the principles of a decision, not just the decision itself. Too often, managers

* https://www.hbr.org/2002/05/read-a-plant-fast

focus on outcomes, but adapting means being able to apply a particular set of principles in a varied way and also knowing how to change the degree of emphasis on a principle, given changing business situations.

One such example is the concept of Lean warehousing. Right away, people run to the 7 Wastes and 5S as principles to apply to a warehouse, but we have to look at the context of Lean and not simply think that all principles in all manners apply to our situation. Let us instead first ask ourselves, of the principles of Lean, what may apply to us and our situation the most in this context of warehousing? Yes, we have the 7 Wastes in our warehouse for sure, but are there some wastes that are more important to remove than others? Do all the wastes apply? Should we change the names of the wastes to motivate, communicate, and allow the warehouse employees to understand the concepts so they will incorporate identifying and removing them into their day-to-day activities?

One such adaptation to the 7 Wastes for warehousing is to focus on Motion, Transportation, and a new way to classify or look at waste: Stagnation. Why should we focus on new terminology such as Stagnation? While Taiichi Ohno stated that Inventory is a waste, by adapting our thinking we can say that the core of this concept was ultimately not that inventory was evil, just that we needed to reduce inventory in order to promote flow. At its core, inventory only occurs as waste because it is stagnant; inventory is not moving toward its customer.

For workers, the question remains how can they reduce Motion and Transportation, while also promoting flow by taking down barriers to Stagnation within the warehouse? This is a good example of having the power of the principles but then contextualizing and adapting it to the situation, while still being aligned to the corporate goals and needs of the customer.

Communicating

Communication is probably the most difficult competency to achieve, yet we tend to spend the least amount of time understanding how to communicate better. A common tool used in Lean businesses is communicating information in a manner that is effective and sophisticated yet exceptionally simple. This is often termed "visual management."

Western culture has not yet learned how to utilize visual communication well, and therefore, we do not have the same understanding of the different layers of complexity that make up exceptional communication. This is because, often, our communications are hidden by layers of other communication, such as work orders in a computer system, handwritten notes, or graphs on a board that cover only one aspect of a process.

Organizational change needs to first begin by understanding and triggering effective communication. This means communication that will cause, or trigger, something to happen. Once this is understood, we must ask ourselves if there is a way we can begin to learn how to sustain this form of communication as the complexity of business increases. It is imperative that communication be learned in this manner; otherwise, the changes that are put in place cannot be sustained.

THE ABILITY TO ASSESS A SITUATION

We often find that businesses don't even possess the capabilities to accurately assess a situation because of a preset cultural bias (acceptable behaviors, known values, and leadership characteristics). These cultural norms predetermine how the organization will absorb, understand, and react to the need for change. When you take this culture into account, of course, any information presented will be viewed from a biased perspective.

Since organizations are made up of individuals, we have to ask ourselves why some businesses are able to anticipate the future and develop amazing products and services. After all, organizations such as Google and Facebook are really not that unique—they began with the same access to resources as every other business. They have the same access to their supply chain, technology, and labor pool as every other business in the United States. So why are they so different? The answer lies in their unique ability to convey the organization's values and the leadership characteristics that affect the behavior of all layers of management required to keep the business running. The difference is their connection between the values and objectives of the organization and those of the individuals within the organization.

Many organizations are able to hide less-than-stellar cultural norms with great innovations, or access to intellectual capital, but rarely does that last over the long term. To get to what makes a company truly successful, you have to look at the day-to-day actions of people and how decisions are made within an organization. To reveal the true values of an organization, you need to begin peeling away the layers of management and artificial business sentiments, such as the mission statement. Look at the core of the

In 2009, when Toyota had a setback in its domination of the automotive industry, Akio Toyoda, the president of Toyota Motors Corporation, directly addressed the issue within Toyota. He did not blame politics, the economy, technology, or specific areas of his business, but instead he went back to the ethos of the business. Mr. Toyoda focused the responsibility on himself and the organization's leadership, stating that they had moved away from the core values that had founded and sustained Toyota for so long. Many people don't know that Toyota came about because of a young boy, Sakichi Toyoda, watching his mother weave fabric for the family's clothing. He noticed how difficult it was and wanted to take the burden of this work away from his mother so she could have a more enjoyable life.

In light of this, Mr. Toyoda has required that Toyota employees once again understand how to become craftsmen and not depend too heavily on machines. What they have found is that automation causes a gap in the knowledge of people. When you rely too heavily on machinery, a widening knowledge gap begins to occur between the work and a human's capabilities to do the work. This leads to complacency and a lack of innovation, because people are no longer intimately involved in producing. Toyota has realized that this need for human interaction with its parts and production process is integral for long-term success.

The stagnation of innovation affected both the strategic outlook and the daily practices of Toyota. It is unthinkable to depend on long-term strategies if your current employees cannot achieve the day-to-day execution of activities needed to meet your stated company goals.

business, the interactions of frontline workers connecting to the business's customers. Once you do this, you can begin to see patterns of actions and reactions within an organization that will show you its ethos, or heart. If you can determine the ethos, you can successfully assess the business.

A number of people who have joined us on our study missions to Japan have come to the realization that businesses in Japan are run very differently from what they are accustomed to. In Japan, company presidents are not just known from afar because of the suits and ties they wear. They are known personally because of their constant presence on the floor, in the field, or in the office. They are well known because they are listening and interacting with their employees, hearing and understanding their hurdles and frustrations—in essence, listening to their customers.

The presidents of these companies are gaining an understanding of their current state and current capabilities in order to understand what needs to change in order to succeed and where they can apply their years of experience to help their customers and their employees. The derivative result of this is that everyone knows them.

In contrast, when we toured a business outside of Japan, the employees did not know who the chief executive officer of the company was (let alone what he looked like) and thought that Collin, who was touring the facility with a keen eye, was their new boss. If an employee who has worked for a company for over 3 years does not know who the boss is, it is safe to assume that they also do not know what the daily purpose and objectives for their job are.

The benefits for managers, owners, and C-Suite executives being known to everyone within the organization go far beyond simply a great knowledge of the operations. The human factor is a more important point, if not the most important, than simply understanding how work needs to be performed.

A great example of an organization that focuses on the human factor of the workplace comes from Collin's publishing business, enna.com. During an afternoon at work, Collin learned that an employee of the company was having an issue with a car dealership. The car he bought turned out to be a lemon. Collin was so infuriated that a dealership would sell such a car, that he gathered together key staff members and they all left at 3 p.m. to protest at the dealership. It wasn't until the police showed up that things got settled and the car was repaired, in entirety, for free.

That story is still shared today. Although the incident occurred many years ago, it is still celebrated as a major event in the company's history to show how everyone looks after each other. That kind of caring goes way further in instilling a sense of pride and efficiency than trying to be efficient every minute of the day.

EFFECTIVENESS: DOING THE RIGHT THING EFFICIENTLY

Too many businesses fall into the trap of focusing on efficiency, rather than effectiveness. Effectiveness is doing the right thing, the degree to which something is successful, with a minimal amount of effort. In Kaizen terms, this would be asking what your customer wants, the demand for your product, and the frequency of orders, then meeting that demand as quickly, or efficiently, as possible. Too often, businesses focus on process-by-process efficiency first, which is a mistake because they need to be able to effectively meet their end-customer's needs, not just process items more efficiently.

To give you an example of a focus on effectiveness, and not just efficiency, let's take a look at Toyota. Most people will be surprised to learn that Toyota does not run a third shift, and a majority of their plants do not run on weekends. From a financial, efficiency, and cost of capital point of view, this is crazy! But it works because Toyota workers are more productive. They have the most reliable systems, the highest quality, and the highest measures of safety in their plants. Even though they are not as efficient, and they do not run their equipment all the time, they are the most profitable. This is what we need to study about Toyota in order to fully appreciate this approach as a strategic and sustainable measure.

To bring this back around to management, we find that not only is there a focus on effectiveness by the workers, but there is also a different prioritization happening with managers in Japan. Japanese managers prioritize their time to observe, test, and verify the true skills of their people, across all levels of the organization. Managers live the values that the business claims it has. Their daily practices are driving the professional values and communication, both internally and externally. The managers take all this in, while also keeping attuned to the changing needs of the market and what the implications of those changes are on their business.

While it is not uncommon to read about the need to walk around and understand your operations and processes, this method does not necessarily lead to success. What is missing from the books and articles that you read is the purpose of the "walking around," or the leadership traits involved in this idea of being present at the place of work. If you do not have an understanding of why you are doing so, then everyone can see there is no sincerity in your purpose of walking around the work area.

A story to keep in mind when you begin your conscious walking around is the story of French writer and illustrator Antoine de Saint-Exupéry. I first heard this story in college and it is still with me. The story illustrates the importance of not jumping to conclusions before proper analysis and highlights the differences in multiple people's viewpoints and focus.

When Antoine de Saint-Exupéry was a child, he drew a picture. When he asked the adults around him if the drawing was scary they all replied, "What is so scary about a hat?" This perplexed Antoine, because in his mind, he had drawn a snake that had eaten an entire elephant (see Figure 1.1).

The illustrator, Antoine, knew what he was drawing, but to the observers, it was not as obvious and what they saw was not what he was intending. This simple analogy highlights the truth in life that intended information does not necessarily carry the same significance or understanding to others. Information needs to hold relevance and be explained and understood from the point of the receiver, not just the person providing the information.

How many times have you been listening during a meeting, thought you understood what the scope of a project was, and then been sideswiped later when you realize that you didn't have a full grasp of the requirements?

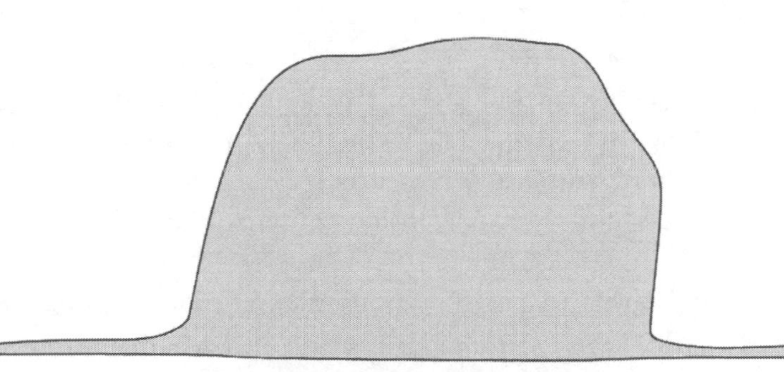

Antoine's drawing shown to adults

What Antoine saw

FIGURE 1.1
Antoine de Saint-Exupéry's drawing.

While some issues can be attributed to scope-creep, most of the time, the root of the issue lies in communication. Though the person doing the explaining often feels he or she is being clear, that clarity doesn't always translate because of additional knowledge they have that others don't, so therefore, the message is not received completely.

In this next story, six blind monks encounter an elephant; the first monk is near the back of the elephant and reaches out, touching the tail, and concludes that they have encountered a rope. The next monk touches a leg, notices its girth, and suggests they have run across a tree. Yet another monk finds the trunk and after feeling down its length, concludes that it is a snake (Figure 1.2).

While all of these observations are correct, each piece individually appears to be something else; all the monks fail to get the full picture. This story illustrates the power of not fully understanding a situation, something that is commonly found in management and leadership within most organizations. All too often, we jump to conclusions based on assumptions, not facts. We do not take the time to refine the tools of analysis and perspective necessary to gain a greater understanding.

The biggest takeaway from this is that there are two basic hurdles that managers must overcome: First, a manager must understand the current state of the business and how it is functioning. Second, one must not be too quick to jump to conclusions of what may be going on in the business.

You need to develop a system that allows you to have a much broader perspective so you can receive input and observations that are beyond the

FIGURE 1.2
Six blind monks.

"perspective as a monk" in our previous example. In all your interactions as a manager here are four key points to always consider:

1. First observations may be deceiving.
   You need the perspective to ask yourself if we are getting to the root cause of the issue.
2. There is a difference between function and performance.
   Try to address issues within the "black box." Make sure you are addressing the functions, not just the outputs or results.
3. Repeated and direct study may be required to insure full understanding.
   You may need to go back to the issue many more times than is convenient to understand the entire problem placed before you.
4. Question whether you are seeing the complete perspective.
   As the examples of the boy's drawing and the blind monks show, seeking the best vantage point is essential to guide others to success.

Toyota states that it is creating an organization that implements the scientific method through all areas, with small experiments at every level. This develops a more sophisticated organization that shares different perspectives to reach the best outcome.

Japanese senseis are most often quoted as telling managers they must first "go to the gemba." This concept cannot be stressed enough, even if it is costly for the visit, because the sensei needs to see the operations in order to see the truth. What we mean by "the truth" is the spirit of real information and real products—the truth is the source. This inhibits misleading conclusions (unintentional or not) and allows for correct decisions to be made from the information.

An example of going to gemba to take a look at the truth of a situation is an operating room in a hospital. While touring a hospital facility, Collin asked to see the cases for the charge nurse that day. She went over to the scheduling board and showed us all the operating rooms and their respective schedules. We were very impressed with their excellent daily management and visual tracking of planned and actual performance.

However, as we started to understand the actual situation, it was revealed that things were not going according to plan. What were the reasons? As it turned out, it was not a matter of how the procedures were implemented or conducted, but more how the patients matched the talent from operating room to operating room. The flow was good when the operations were

There are three aspects to a Skills Matrix:

1. State the skills of your staff
2. Have a plan and process to increase the skills of your staff, and
3. Correlate your staff matrix to client's unique needs.

The first aspect is typically done well, but the last two are usually not considered, although they need to be.

going smoothly and as planned, but when something did not go according to plan, there was a mismatch between specific operation requirements and the skillset of the staff in the operating room.

With multiple rooms operating in parallel, staff allocation is key to smooth and timely surgeries. This hospital had an excellent visual management system, but they also needed a visual skills matrix to help them plan and allocate staff dynamically throughout the day. In fact, as we were visiting, they came to the realization that their real-time schedule was actually working against the skillsets of their staff.

Complex and similar-type operations were happening simultaneously, complicating or even stopping operations due to staffing constraints. By visualizing skills and procedures, they are now able to shuffle operations and level load based on the skillset of their staff on a shift-by-shift basis.

Another observation during our tour was the realization that surgeries were not being classified accurately, and therefore, too many staff were being assigned to simpler procedures and not enough to more complex surgeries. This meant that managers and supervisors were employing more staff than necessary. Their buffer was people, but that didn't mean that they were the correct people for the mix of operations for each day. There was no way that we would have ever seen the reality of the operation with a presentation and meeting outside of the actual operation theatre.

EFFECTIVE OBSERVATION

It is not enough to simply be present at the workplace; an effective manager needs to practice effective observation. So much more than simply watching work being performed, effective observation is the act of being sensitive to the reality of workplace processes. Managers must look at processes with a sense of urgency, not complacency, and a sense of how to make each process a challenge for everyone, or how to make it more interesting to work within. This kind of perspective encourages others to break through their barriers to higher commitment, sense of purpose, and seeing themselves as part of a larger goal.

To become an effective observer, a manager must know, prior to observation, the objectives of the department, process, and workspace. Some questions that can be asked to ascertain these objectives are as follows:

What are the conditions of the day?
How many people are on staff?
How many orders are scheduled to be fulfilled today?
How much work-in-process is there?

These questions must always be answered, regardless of the work area the observer is in, be that on the shop floor, in the design department, or in the office.

Beyond knowing what questions must be asked, the observer must also choose an area to focus on that is based on where they need to strengthen the fundamental values of the organization. For instance, say you find yourself looking at a kitting process for an assembly line. Overall, this is a noncomplex job. Yes, it involves the precise selection of materials, and quality is important, but what is the motivation for the person selecting products for an assembly line? Why should they care? You have to connect the values of the organization to this exact job and role.

At a hydraulic actuator business, this role means selecting parts for the assembly of an actuator. The part picker repeats their process roughly every minute, so it is important for them to know how their job is affecting the end customer. Their job, done correctly, ensures that fluid stays in the actuator so that someone in the world who has a broken gas line, has to repair a water line, or has to extend some plumbing can have his or her project completed on time. These are the connections that need to be made—that this person's job, picking parts for the seals of an actuator, is just as important as the machined gears.

No matter what aspect is being observed, it needs to be assessed with intent. This intensified attention is meant to help team members identify issues that may lay dormant. Managers need to repeat this process of observation over time to get an accurate understanding of the differences between their visits. This understanding of change over time will allow them to get a good sense of the needs of the work area.

We were once told by one of Miura's senseis, Mr. Hitoshi Yamada, that he walked around and observed a company for one entire year before making inroads with the employees and changing the company culture. A manager needs to have patience as well as purpose. Mr. Yamada had patience because his goal was to change the entire business and he needed to achieve this change through observation and building relationships.

In order to get a good baseline understanding of the skills required to effectively observe a workplace, we recommend reading an excellent *Harvard Business Review* paper, "How to Read a Plant Fast," by Eugene Goodson. This paper provides the overall perspective needed to bridge the evaluation knowledge gap for people just learning and acquiring the skills needed to assess an organization.

Mr. Goodson classifies the observations necessary to properly assess and read a plant into 11 categories and provides great insight into valuable criteria to use. As the paper explains each of these categories, Goodson outlines numerous questions that are practical to ask in order to assess the cost of manufacturing or the cost of sales.

Goodson's 11 categories are the following:

1. Customer Satisfaction
2. Safety, Environment, Cleanliness
3. Visual Management Systems
4. Scheduling System
5. Use of Space, Movement of Materials, and Production Line Flow
6. Levels of Inventory and Work-in-Process
7. Teamwork and Motivation
8. Condition and Maintenance of Equipment and Tools
9. Management of Complexity and Variability
10. Supply Chain Integration
11. Commitment to Quality

While there are other observation tools more closely related to Kaizen specifically, we feel that the categories outlined by Goodson are an excellent tool for observation of the current situation of any business.

Having these categories with you as you observe a workplace helps provide a more comprehensive understanding of the business you are observing. You are receiving firsthand information, instead of having to try and be sneaky about gathering facts (for example by asking people questions that may be interpreted in a variety of ways), and you also gain the proper framework to categorize and reflect on your observations. Oftentimes, people are taught to use secretive techniques to observe; however, it is our belief that open and upfront observation, as well as gaining the employees' understanding of the reason for the observation, leads to much better information. This is because the information leads to insight about why the operation is in the condition it is in, while avoiding instilling a sense of fear or mistrust.

Remember, the current state is just a derivative of deeper underlying issues. This is what you are ultimately after, not the symptoms but the underlying facts.

MANAGEMENT BY WALKING AROUND

One phrase commonly used when talking about going to gemba and seeing the process is "Management by Walking Around;" however there is more to management than walking around. You must have a purpose to the walking, you need to have a framework for observation, and you must be able to assess in order to reach a good conclusion. This is the role of managers; the job is more than simple management based on a schedule; it is analysis by using your legs, not just your brain.

Management by Walking Around is a misnomer, as true management actually involves the tools and skills to be impartial and to observe with wise but honest eyes in order to form an objective opinion. This is harder than you think. When we take participants on our study mission to Japan, at the beginning of the week, most people believe they know how to see Kaizen, but most often, they only know one facet of it. However, by the end of the week, they know and understand the ethos of Kaizen: the underlying drivers that make the methodology successful. It is only by seeing various industries and various businesses at different stages of Kaizen promotion that they are able to observe and make a connection to the common thread within them all. Through this exposure to other businesses, the participants are able to then draw upon their own experience and see how this new understanding of Kaizen fits into their own business situation.

The key takeaway from all of this is that we all need to relearn how to see—how to become more sensitive to the reality around us. Our busy schedules impede us from becoming fully immersed in the reality of our businesses, and tools such as Goodson's 11 categories allow for deeper observations and understanding of our workplace.

MANAGEMENT TWO LAYERS DOWN

One very real concern of upper management, or the C-level suite, is how they are able to filter, or qualify, the data they are receiving from their direct reports. This is where managers need to use their skills to determine what is, and what is not, accurate information. It is hard to filter unless you know where the information is coming from. How can you assess something and filter it if you do not have an understanding of where it is coming from, or whose viewpoint it is being seen from?

One great example we have learned by touring so many companies is that each layer of management needs to be able to reach down past their direct reports to the layer below. This is management by two layers. You may manage the layer below you, but you at least need to be observing the layer below that (see Figure 1.3).

The most successful businesses have this two-level management structure. If you don't use this method, it is impossible to properly guide and coach your management team because they don't have visibility to advise or develop at all layers of the organization. This kind of structure should be implemented at all levels of the organization, and a good place to start is with the frontline supervisors, who need to know their team lead's work.

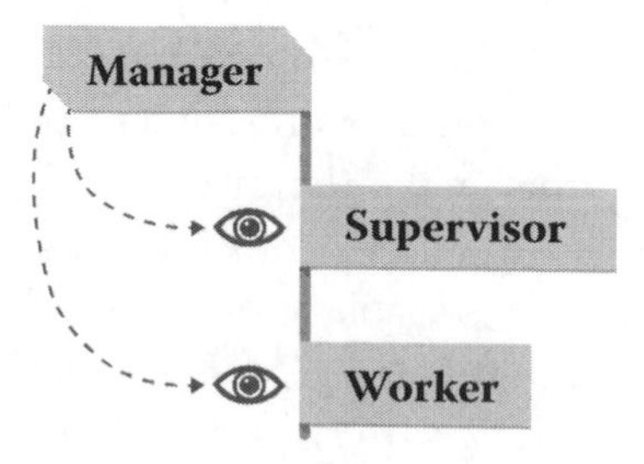

FIGURE 1.3
Management by two layers.

Providing this kind of management principle and structure also conveys a sense of honesty throughout all layers of management, as all know each other's roles and responsibilities in depth. Some may interpret this as snooping, and we understand this is a natural perception, but we have found that once everyone realizes that the true purpose is to allow for more accurate, quicker, and relevant information, they begin to focus on moving forward and not on "protecting" information.

If an organization is set up with full transparency, then the collective energy goes toward other aspects of business, such as thinking of innovative new ways to create value, rather than protecting one's own value. Using this method of transparency leads to layers of innovation and better, more sincere, relationships, both internally and externally to the organization.

STARTING WITH A THEME

In Japanese, they use the phrase "Genba Shugi," meaning to see with your own eyes, but your eyes need to know what to look for. For this, the Japanese first start with a theme, then they consider the right time, and the right depth of analysis for the situation.

Let's take an example of a theme being safety, and the manager of the business in question is considering a time to observe this. Taking a look at the statistics on hand (which were luckily available to the manager), it is determined that first thing in the morning and at the end of the shift are the best times to observe. Why is this? Because humans have a great ability once warmed up to use their dexterity and agility to mask safety issues when everything else is running well. But in the early morning, we are just starting to get moving, and late in the shift, we are tired and therefore are not as dexterous. This presents an opportunity to carefully look at safety, because a business needs to protect its workers when they are most vulnerable.

Taking another example from daily life regarding a theme, I am sure everyone at some point in his or her life has had this same type of experience: You purchase a new car and suddenly you see the same make and model of your new car everywhere.

For Collin, this was most pronounced in the birth of his first child. Suddenly, he and his spouse noticed everywhere they went, there were more babies and cars with those little silhouettes of families in every parking lot. It wasn't that there was a sudden population growth in his hometown, it was that he was more aware of that theme of life because of his own situation. This is why it is so important to have a purpose when observing, so that you don't naturally fall into your own personal bias of observation.

The right attitude is absolutely essential to these observations, as such visits should not be dreaded by people on the floor, but seen as opportunities. The manager needs to communicate that the purpose of his or her observations is to gain a true understanding of the reality of the situation, and that the manager values forthright information and the contribution of individuals. In many ways, the employees have become the customers, inputting information to the manager about what needs to be improved and ways that improvements can be done. In this way, managers can build an emotional connection with employees through shared experience, strengthening their loyalty to the organization and commitment to their work.

Make this walk-through a priority in your schedule so that it is given the proper time and authority to govern your schedule and not be overrun by competing obligations. This is one of the most important ways to connect with the truth of your operations and can influence how changes are made as well as the spread of Kaizen throughout your organization. Many managers may be able to walk through their operations multiple times in a day, and we encourage this in addition to focusing on second- and third-shift operations.

If you are running into issues of quality, long lead times, defects, and issues of customer service, let your employees know that the organization is having these issues. Ask them questions specific to this subject, or theme, and ask for their insight and evidence because they are closest to the work being done. When all activities that the end customer is burdened with are known in the organization, all you need is a purpose, then your employees and you are well on your way to solving the issues.

Assess the benefits of this practice and keep a diary of insights, categorized by business issue. By listening to the perspectives of your employees and getting their input, you will connect them to strategy and customer satisfaction. By giving them input, you are allowing and encouraging them to contribute to success beyond the direct service or product.

CONCLUSION

The reality of your business is a reflection of your employees being given the opportunity to be engaged in the process and connected to the business and its customers. Engagement cannot be forced on anyone—it is something that is given. Leaders set the tone and are responsible for pulling ideas and creating an environment where employees want to apply their discretionary efforts to their work life. This is not required, but when it is achieved, there is more joy in work and more effective productivity as a side benefit.

It takes practice to lead and empower your employees in this way. It will take a different infrastructure to implement their ideas, which you are now actively seeking. It takes time and engagement to get to root causes and improve operations. Such behavior by management helps create a culture that is effective, adaptive, and joyful. Just like a sports team is competitive, demanding, and fun; at work too, we can have an environment that is joyful and productive.

2

The True Meaning of Kaizen

In the first chapter, we explored the complexities of management and the need for observation at the workplace, however, effective management is just part of the key to an engaged workforce. The other key component is the true concept of Kaizen.

No matter where we travel around the world, we always encounter managers who say they are specialists in implementing Lean Manufacturing. Yet when we show them around facilities that either of us have coached, the reaction is nearly always the same:

> I cannot believe that each worker in the business proactively wants to implement Kaizen in their workplace by themselves. I cannot believe you are able to achieve this kind of mindset. How do you do it?

It took Miura a while to accept the fact that what was considered normal in the businesses he coaches in Japan was, in fact, something that other companies around the world desperately desired. Managers outside of Japan seemed to understand the concept of Kaizen as "continuous improvement," but they did not understand the truth behind Kaizen and its original meaning as defined by Taiichi Ohno. In this chapter, we will explain the true meaning of Kaizen and the concepts behind achieving a true Kaizen culture around the world.

Kaizen consists of two Japanese letters, or kanji (Figure 2.1).

改
善

Kaizen

FIGURE 2.1
Kaizen kanji.

Kai means to "change" and Zen means to "be right." Therefore, Kaizen means to change for the better to get closer to the right state.

This has not yet been defined as such in the western world—the concept of the "right state."

There is far more meaning behind the kanji than we are able to articulate outside of the Japanese language itself; however, some of that meaning is that we are supposed to strive to bring things to a more "right state" of being. This is more of a philosophy than a methodology and therefore is a great common thread that can be used throughout the functions and processes of any business. Changes made do not have to be big, as long as they are an improvement. With this mindset of making everything better than the current or past state, we can apply this concept to the individual level (personal) as well as the surrounding environment.

Now let's look at this idea from a business's point of view, and the roles within a business. Organizations put a lot of significance on QCDS, or quality, cost, delivery deadline, and safety. The main goal here is to manufacture products safely, with a high level of quality, and provide customers with these products at lower prices while meeting the promised delivery dates.

With this focus in mind, the true Kaizen culture ought to promote knowledge and ideas generated from everyone involved and must be established from the shop floor in order to follow the requirements given by QCDS.

Consider your daily life in terms of your commitment to your work life. Out of every 24-hour day, where do you spend most of your day? If you ignore the time allotted for sleep, you spend a majority of your waking hours at work. While you are working, are you told to do the same job repeatedly, endlessly, as if you are nothing more than a human component that could be substituted by any other person or machine? If this is the case, is there any satisfaction in your daily work? We would venture a guess that no, there is no satisfaction in this type of work. This sort of mentality makes work an almost torturous task that you must accomplish each day.

With this type of dissatisfaction toward your own work, the level of quality in your output will likely be diminished and it will become extremely challenging to achieve a higher level of productivity.

Workers are not robots. They are human beings and it is essential for everyone involved in business to remember that. Any living organism on Earth, from tiny scrub bushes to the most complex creatures, strive to grow little by little on a daily basis so they can advance a step further in life. This is what it means to truly live your life to the fullest. This is the truth behind Kaizen. It is through Kaizen that you will be able to change the mindset of each individual within a workplace so that everyone says to each other and believes in "Let's make today better than yesterday."

As you can see from this explanation of Kaizen, Kaizen is not limited to simply bettering your work. It applies to everything in your life. Kaizen has often been mistranslated as simply "Continuous Improvement," which explains only one aspect of the truth behind Kaizen. The definition of Kaizen must include the innate desire within each individual to want to make today better than yesterday, so that their quality of life can be significantly enhanced.

DIFFERENCE BETWEEN KAIZEN AND IMPROVEMENT

In 1984, the Toyota Motor Corporation opened the NUMMI Factory (New United Motor Manufacturing, Inc.) in Fremont, California, thanks to a partnership with General Motors. Workers who were designated to work in the NUMMI factory came to Japan to participate in on-the-job training at Toyota's factories, in order to learn about Kaizen. When they returned to the United States, they continued to use the word "Kaizen" because they could not find any English words that articulated the concept. The English word that is closest to describing Kaizen is "Improvement."

However, as we know it, improvement's meaning is to use monetary investments to make things better, whereas Taiichi Ohno's Kaizen is simply the process of implementing the ideas of each individual to make things better. In other words, "Improvement" is to better your goods and materials, and "Kaizen" is to better an individual's actions.*

Let's look at two tales of improvement: the American and Japanese approaches. Americans did not always struggle with factory production. The American use of mass production and automation was one of the key factors that created the worldwide automotive industry. Back then, it was common practice to invest in machines to increase productivity and efficiency, and the result was years of industrial success.

In contrast, the Japanese industry was keeping up with the productivity of the United States by using a completely different approach. In the book published by Taiichi Ohno in 1978, titled *The Toyota Production System*, Mr. Ohno revealed that it was the implementation of the Toyota Production System that allowed Toyota to achieve such high productivity and quality. Toyota did this by switching from a mass production system to a system of producing a high mix of individual product lines (better known as high-mix, low-volume).

When we look back at the Edo Period (1603–1868) in the history of Japan, the art of making things, known as monozukuri, was handled by a handful of highly skilled masters who truly excelled in their craft. They were called "shokunin" (master craftsman). In the manufacturing fields, workers had learned from their masters and through the years of experience working under their mentors they were able to achieve "shokunin" themselves. Without manuals or the ability to replicate that personal training,

* Hitoshi Yamada, *Kaizendamashii no Sakebi*, Nikkan Kogyo Shinbunsha, 2011.

the ability to create and expand the number of skilled laborers was limited. As many Japanese industries suffered from poor financial investment made in the past, their way of Kaizen was not to rapidly increase productivity by spending more money but to utilize each individual's brainpower for creativity and generating new ideas, in order to produce much higher-quality products.

During this time, the Japanese already had a cutting-edge system of recycling, for instance the leftover peels and stems of vegetables were collected and used as fertilizer for farming. Water used for rinsing rice grains prior to cooking was also collected and sprayed across farms to fertilize crops. This recycling system was driven by the desire to make use of things that were normally thrown out, and it was natural for people to have this mindset back then. These ideas did not magically come into being but were a result of applying creativity to everyday situations in order to enrich their lives.

In spite of the fact that Kaizen originated from Japan, people around the world feel a strong connection to the truth behind Kaizen, even though it cannot be easily translated into their language. This is because everyone shares the common desire to use their creativity in order to improve the quality of their day-to-day lives.

To give more personal examples of this innate wanting to make life easier, take a look at your own kitchen or bathroom. Do you place soaps, shampoo bottles, spices, and utensils in more convenient places, so you can quickly reach them? Of course you do, because it makes your cooking or cleaning tasks easier to accomplish. Examples of this can also be found in the hospitality industry. The next time you stay at a hotel, pay attention to the pens that are available to you at check-in, and how they are placed at ease for you to sign your paperwork, oftentimes upright in a penholder so you can more easily grab the pen.

These small ideas, implemented by individuals on a daily basis, in almost a state of subconscious mind, are certainly admirable examples of Kaizen. People do not need to find the right words to define Kaizen in English because they can relate themselves to the truth behind Kaizen, which will allow them to directly accept Kaizen as a new concept in their own environment.

We want to stress again that Kaizen is not all about making your work easier but is about improving the quality of your life. We want to share our passion, through this book, for helping people learn to keep practicing Kaizen for the enrichment of their own lives.

WHO PERFORMS KAIZEN IN YOUR COMPANY?

Kaizen should be executed by all levels of employees, from the president down to the janitor. So we want you to ask yourself: in your company, who is in charge of doing Kaizen? If the only people involved in coming up with improvement ideas are managers, then you are wasting every other person's potential and creativity, which could be used to solve issues. As we talked about earlier, the truth behind Kaizen is that every individual has a strong desire to improve his or her life. So why do we continue to believe that only managers have that desire in business?

Consider the role of employees in your facility. Why do you hire people in the first place? If they are given simple tasks to perform repeatedly all day and are expected to follow their supervisor's orders religiously, they are being utilized as little more than human machines: an infrastructure cost. If you discourage workers from expressing their ideas and treat them like machines, or robots that will always follow your orders, people will become just that. They will lose their innate desire to improve their surroundings.

If managers decide not to let people do Kaizen because they believe that their employees have no ability to implement Kaizen, and people decide not to even try because they are told by their superiors that there is no need for their ideas, there is no way to draw new, innovative, and creative ideas into the business. You will have effectively shut the door on the creative potential to improve your business.

The foundation of Japanese management is trust in people, which encourages each individual to suggest ideas for the purpose of achieving a collective creativity. Miura's Kaizen sensei, Mr. Hitoshi Yamada, learned the Toyota Production System directly from Taiichi Ohno. Thirty-five years ago, he established a leadership-training center called PEC to realize Mr. Ohno's vision to perfect the Toyota Production System. PEC stands for "Personal Education Center for Kaizen" and is designed to draw out the human potential of each trainee through Kaizen. PEC helps organizational leaders create a management system that puts the greatest emphasis on involving each employee's creativity, both in leadership development and problem solving, through the use of Kaizen.

> "One step forward by 100 people is better than 100 steps by a single person."
>
> **Japanese saying**

PEC first coaches trainees by developing their skills in identifying wastes and giving them the encouragement and permission to remove such wastes. By teaching this waste removal concept and the techniques associated with it (Muda-tori) to each employee throughout the organization, leadership (also trained by PEC) becomes able to execute an effective management style that draws out the true potential from each person.

Through PEC, Mr. Yamada has continued to develop the Toyota Production System and has helped tens of thousands of people execute Kaizen by teaching the concept of "Muda-tori." Central to his teachings is the understanding that it is only human beings that can suggest creative ideas, not machines, and that is where you should focus your efforts at work. It is essential, therefore, that everyone understand that employees are not a cost to an organization—they are value-adding agents. As long as employees at every level within an organization, from the highest level down to the shop floor, can implement their creative ideas, the level of customer satisfaction and profitability will increase, while costs are reduced.

True organizational strength is attainable only when every employee can share the same visions and goals and are able to contribute their unique creativity and innovative ideas to problem solving. Mr. Yamada has taught that only management systems that are focused on reaching innate human potential will succeed. While purchased machines suffer dramatic decreases in their capacity as they become out of date, the capacity of each individual can be continuously increased as more training is received.

This understanding goes hand-in-hand with the definition of Kaizen: that Kaizen is not limited to bettering work but is applicable to everything around us. Each human being can do Kaizen, so our goal must be to train each person and get everyone involved so that we see the greatest results. If you trust the true potential of your employees and tap into their brainpower, they will begin adding value to the products and services your organization provides.

When everyone involved reaches this mindset, there will be no limit to how far your organization can grow and succeed into the future. This is not an easy task to accomplish, but it is amazing to behold and it is the goal we should all be striving for.

THE POWER OF 100 IDEAS A YEAR

A pencil and ballpoint pen manufacturing company, Yamagata Mitsubishi Pencil, has been practicing Kaizen under what they call the "2S One-Minute Muda-tori" as a way to establish a solid Kaizen culture throughout their production facility. Instead of the company introducing a Kaizen suggestion system, General Manager Mr. Muto encourages each worker to solve one small problem each day by implementing his or her improvement ideas. An example of this can be as simple as a person putting his tools back where they belong after each use and determining what is and is not useful, daily, in the process. By encouraging workers to implement a small improvement each day, the workers there began habitually sustaining these activities, which is the foundation of a Kaizen culture. The line leaders summarize how many 2S Kaizen improvements each employee implements to calculate the total number of implementations on a daily basis.

In November of one year, they tallied up all of their numbers and found that they had collected 110,000 2S Kaizen implementations. They then divided that by the number of employees (280) and figured they were implementing 393 2S Kaizen ideas per employee per year. To break that down even further, that measures out to an average of 1.6 Kaizen ideas implemented daily, per employee.

Through this process of encouraging each worker to execute small 2S Kaizen ideas, the company was able to establish an organizational culture in which workers started performing 2S religiously and created a strong foundation of true Kaizen culture from the bottom up.

The company also publishes an internal newspaper, called "The Muda-tori Newspaper," on a regular basis so that workers can learn about Kaizen ideas from outside their department (Figure 2.2). They even post the newspaper on the walls in the bathroom so that workers going in and out of

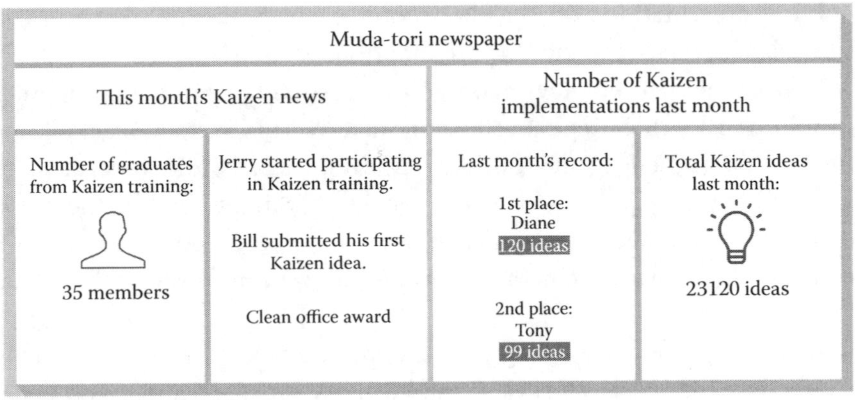

Muda-tori newspaper			
This month's Kaizen news		Number of Kaizen implementations last month	
Number of graduates from Kaizen training: 35 members	Jerry started participating in Kaizen training. Bill submitted his first Kaizen idea. Clean office award	Last month's record: 1st place: Diane 120 ideas 2nd place: Tony 99 ideas	Total Kaizen ideas last month: 23120 ideas

FIGURE 2.2
Muda-tori newspaper example.

the facilities can be exposed to the information. This way of sharing the information in the bathroom was a Kaizen idea itself.

THE CONTAGIOUS POWER OF KAIZEN

A confectionery company, Fraicheur, has been coached by Mr. Yamada for several years and has accomplished some amazing feats. Mr. Yamada visits the company once a month to conduct intensive Kaizen training events. After each training event, the employees from various parts of the factory give presentations to each other and to Mr. Yamada. These presentations happen the same day as they complete their Kaizen efforts.

In the early period of Fraicheur's Kaizen implementation journey, such presentations were given by managers; however, the employees who started proactively implementing Kaizen in the workplace quickly began to give the presentations. They soon moved the presentations to the shop floor to enable better illustration of the improvements that had been made.

The president of Fraicheur, Mr. Hiroyuki Yoshida, has always participated in the presentations on the shop floor. He says that this unique way of showcasing workers who have implemented great Kaizen ideas is extremely inspiring to the other workers. Even new hires, often having worked in the business for only a month, quickly began expressing their desire to do Kaizen and contribute to the team. Not long after they began holding presentations on the shop floor, even their part-time workers started volunteering to give the presentations.

Today, Fraicheur employs many contract workers from the Philippines in addition to their regular staff. Because of this history of employee interaction and engagement in Kaizen, even these contract workers are inspired to participate and suggest new ideas. Surrounded by the president, senior managers, and colleagues from the shop floor, these workers can proudly explain their Kaizen results to everyone, even with limited Japanese.

When a presentation is given, the production line is temporarily stopped so that every worker in the business can participant and listen to the improvements made. After each presentation, everyone appreciates and congratulates the implementers on the positive results of their unique Kaizen ideas. This builds the confidence of the employees involved and increases their commitment to Kaizen.

Fraicheur is another great example of promoting full participation of staff and establishing a true Kaizen culture, where top management and employees are collectively working toward the same goal through the spirit of true Kaizen.

INSPIRATION DIRECTLY INTO PRACTICE ON THE SHOP FLOOR

Kaizen isn't about thinking or analyzing. It is about executing your ideas to change everything to make it better than before, even if it takes baby steps to get there. This means that your actions need to be put into practice wherever work is taking place. It is not Kaizen when ideas are discussed only in a meeting room and some forecasted results are presented as potential answers. Kaizen ideas without any measurable results are idle thought experiments, not true improvement.

As we established in the first chapter, a manager's place is on the shop floor. This is the first step in implementing Kaizen. Mr. Yamada defines a

> Kaizen only becomes Kaizen when results can be obtained by the idea's execution.

> A "shop floor" is the place where a problem is occurring, or has occurred.

shop floor to be the place where the nature of a problem is occurring or has occurred.* When you step onto the shop floor, you will quickly be able to obtain information that is impossible to capture in a meeting room.

Many managers have considered outsourcing their production or relocating their entire production line to a different country, in order to reduce costs. However, by going to the shop floor and observing the work being done, they can easily solve the financial problems they are trying to fix without outsourcing.

It is not enough to simply complete a Lean workshop or training session. You must set up a mechanism for sustainment. If, after a training workshop, production continues to generate defects, this means that items still have to be sorted and set in order, and managers will have to reassess the level of implementation in the business. A cluttered workplace is simply a representation of workers who are not completely engaged and committed to performing their daily tasks well. If employees cannot follow simple rules regarding the organization of their workplace, there is no way these workers can produce high-quality products safely and in a cost-effective manner.

If you want to improve your workplace, physically go to the workplace and begin working with employees to suggest ideas to solve issues. Most importantly, when you come up with an idea, don't hesitate to immediately go and execute the idea. Try it and see how it works!

THE PROCESS OF KAIZEN

The process of Kaizen implementation and improvement can be summed up as follows:

1. Visit the shop floor and observe.
2. Find wastes.
3. Execute the Kaizen ideas you believe will work.

---

* Hitoshi Yamada, *Hitome de Wakaru, Sugu ni Ikaseru, Kiso kara Wakaru Kaizen Leader Yosei Koza*, Nikkan Kogyo Shinbunsha, 2008.

4. Reflect upon your results. (If your ideas did not work, go back to step 3 and begin looking for other solutions.)
5. Find the next Kaizen idea.

Kaizen ideas that are planned without involving the opinions of the employees in the business are likely to fail completely or, if implemented, will need to be revised later. True Kaizen cannot be achieved without the involvement of every worker from an area. The shop floor bridges the gap between managers and workers. It connects everyone's brainpower together to achieve far greater results.

The accumulation of small improvements, executed repeatedly, will lead to significant achievements and organizational transformation over the long run. The more work capacity that the shop floor gains through Kaizen, the more strength it gives to the overall business strategy. Your company's overall financial statement is the evidence to determine how well Kaizen is being applied not only to production, but also areas such as sales, purchasing, and R&D.

CONCLUSION

The true meaning of Kaizen is to empower people to improve all areas of their lives, to become inspired, and to implement those ideas immediately into their surroundings (not just their work). This is the leverage point of Kaizen—it is sustainable and scalable because it is a personal commitment and everyone has the resources to implement their inspirations.

Enough theory though. In the next chapters, we will dive right into the shop floor and explore what can be done, from the ground up, to inspire a Kaizen culture!

3

Monozukuri

While Chapter 2 explained the basis of a true Kaizen culture, the question still remains: what is the best way to create such a culture? This sort of culture does not simply emerge; it is the result of clear and conscious decisions by leadership to change the workplace environment.

It is no secret that the environment in which we live influences our behavior. How we are instructed, relationships between colleagues and supervisors or managers, company policies, and work procedures and regulations have proven to be very influential in an organization's overall performance. As a leader, you need to navigate through these issues to achieve a true Kaizen culture.

THE ART OF MAKING: MONOZUKURI

All too often, the idea of working is communicated as less valuable than other aspects of life. The weekend is the "best time," and time spent at work is just nonweekend time. Isn't it possible to also enjoy work and get a sense of fulfillment and contribution to society?

One of the foundations to engagement and creativity is the concept of Monozukuri. Monozukuri is a term that was used well before modern industrialization, when people had the skills to envision something and then turn that vision into reality. This is the concept known as, and referred to in English as, being a "craftsman."

While the literal translation of Monozukuri is "making things" (mono: parts, objects, things; zukuri: making), the spirit behind the word holds so much more meaning. It encompasses all the skills necessary to sustain oneself, as well as the ability to create something from nothing, for the benefit of the customer.

It also has meaning in the concept of profit itself. The idea of profit in old Japanese is "reward for good service from the customer," not payment or cash received. It means that the customer has shown appreciation. Therefore, the concept of Monozukuri means to have all of the skills and abilities necessary to create something for a customer and be rewarded by the customer's appreciation.

This concept of creating something that is rewarding for the worker, and appreciated by the customer, is a concept that we need to build into all jobs and functions of an organization. We need to ensure we are training and cultivating knowledge and skills in our employees and providing them enough scope of work that they can see their ideas and efforts being appreciated directly from customers.

The idea of Monozukuri is having the ability to own your craft as a professional and continue to change with the times, improve, and innovate your skillset, not simply for yourself but for the customers you are providing value to.

Two organizations that we have toured in Japan that are excellent examples of Monozukuri are Yamaha Pianos and Toyota. All too often on our training missions to Japan, the participants find the concept of Monozukuri counterintuitive, as it brings back the idea of craftsmanship while having the efficiencies of industrial methods. We see the underlying success in the values and principles that each company represents in their respective industry. We think that the best organizations find the set of values that stand the test of time, and these values and principles serve as the enduring success.

Let us first take a look at Yamaha Pianos. Its founder, Mr. Torakusu Yamaha, was a pioneer in production of Western musical instruments in Japan. He wanted first and foremost to instill passion into the organization. This can be seen in Yamaha's slogan, "sharing passion and performance." The Japanese word for passion is "kando": "kan" meaning emotions, and "do" meaning movement. This meaning perfectly encapsulates putting your values into action. Therefore, "sharing passion and performance" sit at the center of the five values of Yamaha's philosophy:

- Customer Experience: joy, beauty, confidence, discovery.
- Virtuous Actions: Embrace your will. Stand on integrity. Be proactive. Go beyond the limits. Stick to the goals.
- Sharing Passion & Performance
- Corporate Mission: "With our unique expertise and sensibilities, gained from our devotion to sound and music, we are committed to creating excitement and cultural inspiration together with people around the world."
- Quality of Products: excellence, authenticity, innovation.

These values are central to Yamaha and the deeper meaning of Monozukuri. By clearly outlining their commitment to their customers, their codes of proper conduct, and how this drives their success in achieving their mission as an organization, they also define the types of businesses they need to partner with (those with similar values).

Similarly, the Toyota Corporation focuses on the value it provides to customers. This is shown in its publication "The Toyota Way," which emphasizes their main focus of "Respect for People" and "Continuous Improvement."

These core values are then expanded into five key elements to more completely explain Toyota's philosophy:

- Continuously challenge the status quo in every element of business
- Respect all parties in business
- Teamwork is the only way to achieve success and shows respect
- Continuous improvement
- Root cause fact-finding on the gemba

The concept of having a holistic approach that involves every person in the organization, and conditions that govern success beyond the balance sheet or income statement, parallel the concept of Monozukuri.

The core commonalities that link both businesses and Monozukuri seem to be the following:

1. Focus on the customer.
   Famous management teacher Dr. Peter Drucker once stated that a business exists to benefit society and the employee.

2. Respect for one's self and others.
   This has deep meaning as it reflects the need to have self-discipline in regard to one's self and, therefore, to society, the environment, customers, and other employees.
3. Always have a sense to challenge the way things are.
   Never be satisfied with the status quo.
4. Continuously improve everything.
   Everything is changing; a lack of daily progress cannot be tolerated.

Looking at these four principles, it is easy to see that they affect all aspects of a business. They are not department based, but are the roots of the organization's behaviors, values, and leadership characteristics. As well, these principles transcend industry, global location, and functionality.

The benefits of these principles allow employees to be guided by the types of partnerships they need to seek, as well as which relationships they need to end. This focus attracts the right customers to the business as well, which in turn benefits everyone, from the community, to employees, owners, and even the environment. Striving to achieve these principles daily allows for more prosperity across the globe.

This point is highlighted when we look back at times of economic recession. We see that when something is not balanced, and principles are driven for reasons outside of a sustainable framework, even a small segment of the economy will drag down the rest. By defining a business's ultimate purpose and values, which are akin to the concept of Monozukuri, the resilience and integrity of the organization will be sustainable, no matter what the economic conditions around it are.

EMPLOYEES' COLLECTIVE MINDSET

We find that the best businesses connect strategy to their employees' daily work. Their employees are more satisfied and create better products and services. These aspects cannot be objectively measured, but they are still important. Quality of management, corporate values, behaviors, and organizational design will lead to success more than only focusing on the numbers.

When all is said and done, the key difference between a good company and a great company are the employees. Even looking at more modern giants (at the time of this writing) such as Apple, Google, or Facebook, we see they have clear objectives and their employees are making the biggest difference in executing the company's strategy.

But what about the financial success of those employees? Are they more successful in this respect? Yes, they are. They have fewer issues with management and attendance. They have proactive relationships, have a joyful interdependency dictated by the company's guiding principles, and enjoy great financial stability and growth. Without a doubt, the environment in which they work is directly connected to the financial benefits they receive, but the financial benefits are a derivative of great culture. When we say environment, we mean the social environment, not just the physical environment. Some businesses, such as construction or repair, cannot control the physical environment where work is performed. In this case, the core environment is the social environment and values that guide interactions on the job.

As an example, Gallup has studied employees for decades in over 160 countries. Their depth of knowledge about industries, cultures, and regional nuances is nearly unparalleled, and through their years of knowledge, they have found that there are 12 things an employee needs in order to succeed at their job (and subsequently for the business to succeed as well).

These 12 Factors, from the employee's perspective (Figure 3.1), are as follows:

- I know what is expected of me at work.
- I have the materials and equipment I need to do my work right.
- At work, I have the opportunity to do what I do best every day.
- In the last seven days I have received recognition or praise for doing good work.
- My supervisor, or someone at work, seems to care about me as a person.
- There is someone at work who encourages my development.
- At work, my opinions seem to count.
- The mission or purpose of my company makes me feel my job is important.
- My associates or fellow employees are committed to doing quality work.

Gallup's 12 Elements of Great Managing
I know what is expected of me at work.
I have the materials and the equipment I need to do my work right.
At work, I have the opportunity to do what I do best everyday.
In the last seven days, I have received recognition for praise of doing good work.
My supervisor, or someone at work, seems to care about me as a person.
There is someone at work who encourages my development.
At work, my opinions seem to count.
The mission or purpose of my company makes me feel my job is important.
My associates or fellow employees are committed to doing quality work.
I have a best friend at work.
In the last six months, someone at work has talked to me about my progress.
This last year, I have had opportunities at work to learn to grow.

FIGURE 3.1
Gallup's 12 Factors of Management.

- I have a best friend at work.
- In the last six months someone at work has talked to me about my progress.
- This last year, I have had opportunities at work to learn and grow.

Clearly, we are not talking about daily production goals or defect rates here. We are talking about factors that affect the emotion of employees. These are aspects that they find important in order to go about their life, career, and daily routines. As a manager, you need to deal with the day-to-day demands of activities, but you must put more weight into the design of your social structures in order to support the people who work for you. You need to meet the daily demands of work activities, but also ask yourself how you are supporting the emotional and social needs that drive us all.

We know that this is an ongoing issue for everyone because we have three decades worth of data to look back upon, and we find that Gallup and other similar organizations that measure culture are still showing similar results as they were years ago (see Figure 3.2).

With the reality of the work climate, what it continues to be, and the general health and sense of purpose of employees being so low, what we are experiencing is a failing grade for organizations as a whole. To overcome

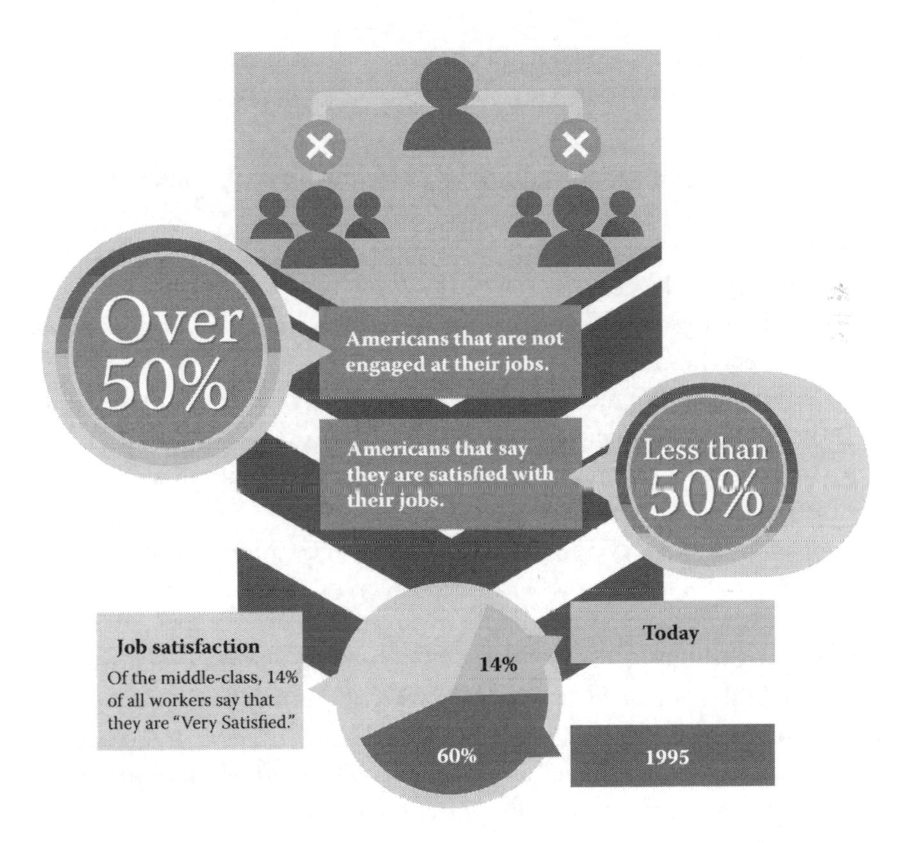

FIGURE 3.2
American job satisfaction statistics.

these statistical truths, we have to understand what motivates employees to be engaged in their work. This is a powerful concept, as it is an emotional commitment that cannot be easily seen, nor replicated. The definition for engaged is, "an employee's willingness to put their discretionary effort into their work life."

When looking at the businesses we are invited into, we find that, although they are in highly competitive markets, they are able to succeed without the best funding, the best technology, or even the most modern methods. What sets them apart are the many principles and designs that allow them to win when competing for the customer.

They are striving to create a more successful corporate environment by empowering everyone from the frontline employee to the board of directors. What do they focus on to create this Kaizen and Monozukuri environment? The best companies that we have toured have shown us their formula of success and they have very common themes:

- Values principles and mindset: Every employee is aligned to make decisions respectfully.
- Mission and purpose: They are simple and sophisticated to connect with all employees.
- Challenging environment: Employees and managers know which skillsets they need to improve and invest in.
- Personal development: Employees can clearly see a path to the future development of their careers.
- Contextualization: Managers and leaders contextualize corporate goals to employees' daily work.
- Recognition every 3 weeks: This cycle of recognition of achievement has the best mix of frequency and success.

If you reread the 12 Factors that Gallup outlines through its research, you will find a lot of focus centered on community and the soft-skill aspects of management that are rarely talked about in most organizations. We personally like this focus, as we have seen numerous times how it has been used to great effect in a wide variety of companies.

Getting to these factors is not easy, and we need to understand the history of how we got to where we are now. Since the beginning of management principles, science, and the modernization of business techniques, there has been great success in applying rational and scientific approaches to management.

If we go back to the beginning of the industrial revolution, there was a need for a measured approach. For this reason, people focused on the processes and not the people factor. Looking back on the evolution of management, it started with a very uneducated work force. Dependence on, and the leverage of, the machine was the greatest focus of businesses, in order to ensure consistent and productive outcomes.

Times have changed in the twenty-first century. Even today, computer programming is becoming a commodity. So, what is left to do? We have such a well-educated workforce and so much competition that we have to consider how to motivate, drive, and provide purpose to our employees in order to maintain and grow them in our organizations. The management of the future will be how to leverage people's creativity by providing them with an environment and challenges that stimulate further purpose than simply a wage or good working hours.

Management needs to change. Our workforce is truly the most valuable asset, and machines are the least expensive resource to acquire now. The organizations that address the more intuitive, emotional side of employees will be the successful organizations of the future (Figure 3.3).

This is where you find the emotional and creative dimensions of people's psyche. Both are absolutely necessary for effective management strategies (Figure 3.4).

It is necessary to consider this emotional dimension because of its impact on the workplace. Look back over the list of Gallup's 12 Factors showing employee's satisfaction. How many of those statements were qualitative (i.e., feelings) rather than hard data? While these factors are specific, the actual measurement and assessment of these dimensions are complicated simply because it is based on perception. People will respond differently in any given situation, so leadership cannot be standardized. Where the best businesses will not have a standard that is forced, they will abide by principles, behaviors, and values, regardless of the situation.

Many books promote "Leadership Standard Work," but in all the businesses we've seen, true leadership means coaching by principles and

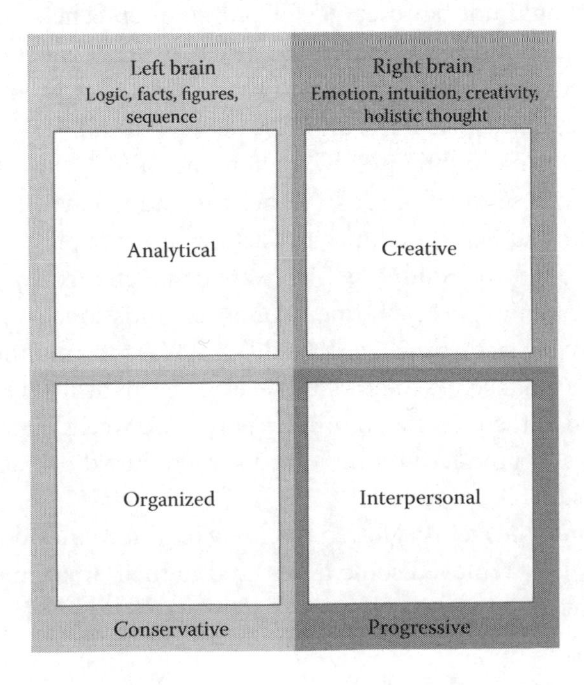

FIGURE 3.3
Leadership beyond the logical left-half of the brain.

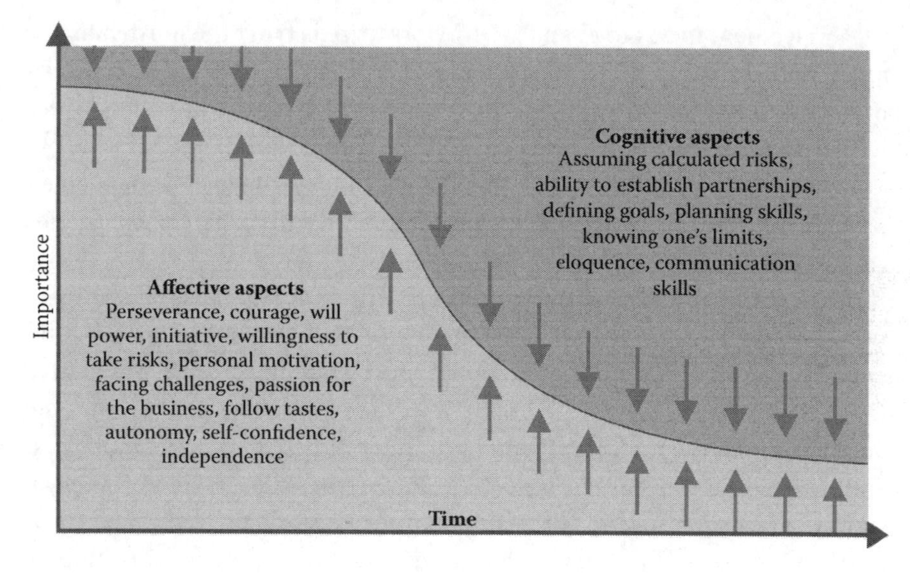

FIGURE 3.4
Cognitive and affective factors.

guiding people while customizing to their specific needs. This kind of leadership should not be reserved for only the top echelons. Leadership needs to be in all jobs, at all levels, in order to be the most effective. We need to respond to people's rational needs, logical needs, and emotional needs and design our system with these needs in mind.

These principles, highlighted by Gallup and others, show the need for people to have a sense of purpose, to belong, and to have others be interested in their well-being. Gallup's 12 Factors, or Principles, are very similar to the Hawthorne Studies, which were conducted many years ago in the previous century. Hawthorne Works* commissioned researchers to assess if physical factors in a work facility, such as the amount of lighting, affected performance. So they added several new lights to the factory and were amazed at the increase in productivity! Two weeks later, they added more lights and productivity increased again. They did it again and got the same result.

At this point, the facility glowed intensely bright, so they decided to test the reverse. They removed some lights, and to their amazement, productivity continued to increase. The researchers were confused, why was this happening?

* https://en.wikipedia.org/wiki/Hawthorne_effect

The answer lay not in the lights themselves, but with the workers' awareness that management was paying attention to them. The reason for the increase in productivity was not a result of the physical areas of the workplace being changed but was affected by the emotional dimensions. The workers, aware that they were being monitored, worked harder every time something changed because they knew they were being watched.

This research was then expanded to explore what factors contribute to productivity. Hawthorne identified that the nature of interaction with the supervisors, being given feedback on performance, a sense of identity in being part of a select group, and a sense of accomplishment all affected the productivity of the workers.

Herzberg's studies on organizations, which would come to be known as the two-factor theory,* came to a remarkably similar conclusion, but Frederick Herzberg grouped the findings into two categories: satisfiers, which are motivating, and dissatisfiers, which are demotivating (see Figure 3.5). Major sources of dissatisfaction were company policy and administrative practices, supervision, interpersonal relationships, working conditions, and salary. Major satisfiers were achievement, intrinsic interest in the work, responsibility, and advancement.

What these dimensions are really driving at is the idea of employee engagement. As we addressed in the last chapter, tapping into the creative potential of the human mind, the true meaning of Kaizen, is essential to creating that sort of engagement.

Corporate-wide creativity depends on the ideas of employees at each level of the organization. Today's new businesses are being created by employees who have left larger corporations in order to fulfill their need for engagement and purpose. They do this because of the possibility of finding new ideas and new improvements. The challenge for managers is to create these possibilities and opportunities for their employees within their organization, in order to not lose them to more engaging opportunities elsewhere.

A recent article in the *Wall Street Journal*† cited that 56% of innovation will come from employees, not managers or leaders. This is the reality that needs to be addressed by current and future managers. We need to unlock the potential of our well-educated employee base.

---

* http://en.wikipedia.org/wiki/Frederick_Herzberg

† http://www.wsj.com/articles/SB10001424127887324178904578340071261396666

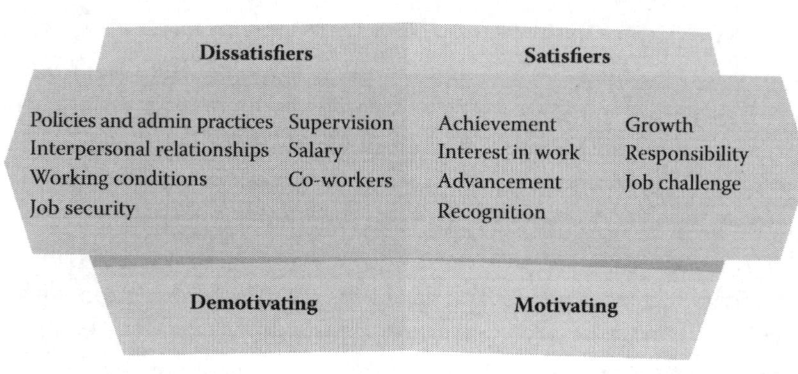

FIGURE 3.5
Herzberg's motivating findings.

Still, the question remains, how do you do this? How do you mold the physical and organizational aspects of your work environment, so that it inspires creative and engaged workers? It all starts with an alluring work environment.

WORK ENVIRONMENT AFFECTS PERFORMANCE

To get the best performance from people, you need an environment that is motivating, encouraging, and matches your values and purpose. Many people think that they need to move into a new building to feel comfortable, innovative, and open to change. But businesses often can't afford to do this. They must modify what they have. Many of the most successful new businesses started in warehouses or old buildings, yet they have an atmosphere that other businesses lack. The location and environment are modified to match the values and purpose of the business.

Every year, business magazines publish special editions with titles like "The 20 Best Companies to Work for," but while their choices are always debatable, it does convey a sense of worker's satisfaction in response to realities found at their workplace. These companies must be doing something to appeal to employees' perceptions or tastes.

Organizations must appeal to a multitude of interested parties. They need to provide the right resources so that employees are able to create the right environment based on the principles, methods, values, and beliefs of the organization. In 2014, the Toyota Motor Corporation's website posted,

"[Toyota] listen[s] carefully to our customers and the local community as we pursue a business that works toward harmony with people, society, and the global environment, as well as the realization of a sustainable society through monzukuri (manufacturing)." This perspective is all-inclusive, holistic, and not confrontational to meet various needs everywhere.

It is important to appreciate the nuances of an operation that contribute to the creation, and then maintenance, of an appealing work environment. It is difficult to measure the benefits of such effort; however, the real benefits to employees, customers, and organizations are undeniable.

Theaters, coffee shops like Starbucks, and even Facebook invest in work environments that convey their principles and values. Employees are more successful if the environment represents the values they are a part of. We have seen organizations focus on the facility itself, the geographical location of the facility, showcasing the organization's values, standards of communication, and the feedback and escalation process within the day-to-day business.

There are two very well-known studies of behavior that consider the general environment and the individual: Fixing Broken Windows by Kelling and Wilson (1982)* and the Stanford Prison Experiment.†

The Broken Windows experiment relates to the fact that if a window is broken, this leads to more broken windows, which leads to graffiti and creates an atmosphere that is conducive to criminal behavior. Kelling and Wilson argued that by addressing quality-of-life issues in crimes such as public drunkenness and aggressive panhandling, versus just looking after serious crimes (such as murder), this would deter more serious crimes from occurring. This also meant that more good families and residents would remain in the towns and neighborhoods.

This theory was applied to the New York City public transit, with the intent to make the atmosphere look and feel safer for residents. The experiment was a resounding success. Crime rates went down and people responded with behavior to match their environmental expectations.

The Stanford Prison Experiment, sometimes referred to as the Zimbardo Experiment, was conducted with 24 college students from the United States and Canada who lived in the Stanford area at the time and wanted to earn $15 a day by participating in the experiment. The experimenters randomly assigned the participants a role as prisoner or prison guard. The

---

* http://www.theatlantic.com/magazine/archive/1982/03/broken-windows/304465/

† http://www.prisonexp.org/

experiment was conducted in a simulated jail that had been constructed on campus.

What was so remarkable about the experiment was how quickly the participants assumed their roles. The prisoners acted against the guards. The guards used force and issued punishment and abuses, even though they all knew going into the experiment that it was just a simulation. The situation degraded so quickly that they ended the experiment early, and Dr. Zimbardo and his colleagues concluded that the situation and environment we are in truly does dictate our behavior.

Bringing this back to the challenges of leadership and the need to develop the sensitivity to observe your workplace, as well as how to incorporate management practices that support a sustainable culture, we have come up with five keystones to creating an environment to promote the best behaviors, values, and leadership characteristics in your employees:

1. Invest and motivate your frontline supervisors.
2. Invest in a workplace that speaks back to the employee.
3. Use One Level Down management practices.
4. Everyone should be a part of the work area.
5. Create a business where communication is free flowing without any fear.

IT'S ALL ABOUT TRUST

Organizations are socioeconomic systems; they are living and breathing entities that require a dynamic approach to change. In organizations, many things are happening simultaneously, both from a social aspect and a process aspect. You have people working together in relationships that might be fleeting at best, due to the nature of the business, and you have other relationships that are consistent and always building. There are functions that may not interact with certain other functions, such as marketing and shipping.

So it is not lightly that we look at how to conduct change within an organization, since it has to factor in all of these aspects. Moreover, our goal is to have all these socioeconomic entities survive and thrive over time. How you get this to happen comes down to the quality of relationships between the people who work together.

Do you trust your boss and colleagues? It is this trust that needs to be met in order for people to share, cooperate, and collaborate together in a manner that is better for all. If you do not have an environment that is trusting, then people begin to act on fear, and fear is a selfish and survival-based mechanism that stops communication and cooperation from happening.

As improvement coaches and publishers of training material around productivity methodologies, we are often met with fear when we go into organizations, regardless of what we say we do, or how the relationship with the organization began. This is because people immediately go to the worst-case scenario and assume we represent a cost-cutting measure. They begin asking themselves, "How will this affect me?" and "How will this affect my job and salary needs?" These are real concerns, and not to be taken lightly. People naturally think of change as a negative thing. It is human nature to resist change because it is unknown and therefore can be fear inducing.

There are many surveys that show that there is a gap in trust between employees and the organizations they work for, as well as a gap in trust, and inherent fragility of relationships, between employees and each of the levels of management. As for trusting the organization itself, Edelman posted their results of their Annual Global Study* (their Trust Barometer), and the results were not encouraging. The results showed that employees "trust a great deal" only 16% of the time when it comes to their organization.

> In regard to their employer, employees say they "trust a great deal" only 16% of the time.

Herein lies the problem. People will not make decisions on their own, or take directions, if they do not trust the company or management. This is an important aspect for management to think about as they attempt to instill change; most organizational climates cannot handle change because of internal issues, not because of a lack of talent or capabilities. Managers must be aware of this in order to align their intent so they can lead change through the not-always-logical landscape of a business.

It is natural for people to resist change. Why should employees make a decision, or do something different, if they do not feel safe or if they feel

* http://www.edelman.com/p/6-a-m/2014-edelman-trust-barometer/

that management will not support the changes they make? All too often, our predecessors have not managed the change process well enough, so current managers do not have the credibility in their position to encourage change in their employees.

If there is not a sincere connection of trust between colleagues and managers before you introduce change, then you already know the outcome of the change efforts. It will work for a while, but as soon as you pull supervision off the effort, it will slowly lose momentum and begin to slide back to the way things were before. So how do you effectively instill change?

As Miura's sensei, Mr. Yamada, is fond of saying, you must build trust before you begin to make any changes. It takes time for people to build up a relationship enough to start looking beyond the role of the manager and see the person that is asking for change to take place. They will see your commitment, they will see your day-to-day actions, and they will always be watching and judging your words against your actions. Because of this, employees need to see integrity in your words and actions before they will be open to change. And they will only be open enough to hear your ideas after you have listened to them and considered their ideas, needs, and frustrations.

Many articles, studies, and blogs indicate that businesses have to deal with soft issues and not just technical requirements. Context and relationships are just as important as the tools. The soft issues of business such as trust take time. Trust is given, not taken, and only time will tell the story. People who stayed at the business through layoffs, management changes, and reorganization have long memories and justifiably so.

Since building trust develops over time, how are you listening to your employees today? How are you visiting the workplace, or gemba, and for what purpose? Why are you making time for an employee? These are all questions that need to be answered, if not directly to your employees then through your commitment to understanding the real situations and getting real information. You must also commit to helping change the organization from the bottom–up and top–down at the same time.

Another important aspect is to truly hear your employees' concerns and take them to heart. You must ask people the how and why of situations, as well as possible solutions they might have. You must also communicate the rationale for decisions and situations that need to be addressed, providing individuals with better understanding of the pressures the organization as a whole is facing. While it can be tempting to simply issue a command, spending the time to explain how and why you go about thinking and making the decisions you do is the most valuable aspect of your interactions with employees.

CONCLUSION

Providing your thinking behind decisions not only goes a long way toward building trust, it can even encourage good decision-making on the part of employees because they have more knowledge and information. This is the way to build trust and coach your employees.

Similarly, employees who feel involved in decisions that affect their well-being are more likely to perceive their leaders as acting with integrity. To motivate and engage with employees, it is essential for managers to tap into the emotional side of their employees. The environment people inhabit affects their behavior. By building trust and creating an attractive workplace, managers can motivate and inspire their employees to produce real results.

4

Communicating the Workplace Culture

What are the magical elements of culture that we see in the top businesses of the world? We see organizations with extremely short value statements that resonate and appear to mean so much to people. Google's mission statement is a prime example: Do No Evil. How does that mean anything to today's employee? It might be difficult to quantify, but it seems to work from a performance point of view.

Values and principles are the building blocks of corporations, and having a sense of purpose is key. These are the elements that hold people together, the glue that holds people's mindsets and keeps them working together well. We believe this is one of the main reasons why methodologies such as Kaizen and Six Sigma work so well in corporations, as they provide a common understanding so people can work more effectively together.

So what does this have to do with culture? Culture is the learned, socially acquired traditions of people within an organization. Values speak to traditions, so people learn what is socially and professionally acceptable at an organization, and this drives their thinking, behavior, and patterns of interaction. This perpetuates upon itself until it is infused throughout the organization, at all levels, and we then refer to it as the "company culture."

We express the culture within an organization by how an organization is structured, how they educate and train their employees, how they recruit talent, and how they hire. The only way to have a concise and meaningful overall process is for each of these individual processes to work together, based on a common theme and belief system. These kinds of beliefs are then reinforced through an employee review process, promotions, and of course, pay raises.

As we have seen on my benchmarking trips around the world, the culture of an organization is present even if it isn't visible. Michael Watkins* says that "culture is the organization's immune system." From this perspective, culture is not only a learned process but also an innate organic process. It is there to protect the organization and to look after problems and opportunities. A more classical view is that of John Kotter,† who places culture as a result of management and organizational processes—a top-down process of planning and executing culture by upper management.

Since social media has turned every single employee into a potential advertiser for the values of your organization, it is essential that you build, and most importantly, maintain a positive organizational culture so that every time employees step up to their digital bullhorn they are broadcasting a positive message about the place they work.

We have seen that culture can be divided into two components: values and performance. Values are believed to be less tangible, but there is a massive amount of evidence that through good values you will see higher performance results. Gallup shows that it is the cultural elements that make the difference between the best corporations, and those that are just average. The average return for corporations with high culture and engagement is 50% higher profitability than those without. If you look up corporate profitability based on performance alone, without seeing it as a derivative result, this would be a mistake. The goal is not performance, but an organization that connects its people to the customers who shows appreciation for that connection in the form of purchasing the company's products and/or services.

Values come first, and then performance in the form of financial measures, such as higher profitability, come about as a result of aligned and delivered values that connect with the customer.

So how do you set the right values and create a positive organizational culture? It starts with developing a way to project the ethos of your organization.

One easy way to share elements of the ethos of your organization is through using visual displays to convey information in a meaningful way. Visuals are more easily remembered than words alone. We have seen no better example of the layered effect of visuals than on our benchmarking trips to Japan. The Japanese management systems are profound in their ability to layer values and information into visual communication at all levels of the organization.

---

* https://hbr.org/2013/05/what-is-organizational-culture

† http://en.wikipedia.org/wiki/John_Kotter

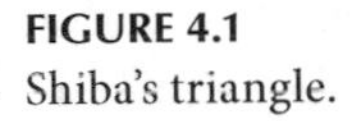
FIGURE 4.1
Shiba's triangle.

One of the best pieces of information we have encountered is Professor Shoji Shiba's Triangle. Professor Shiba is an internationally renowned expert in Total Quality Management, or TQM. He illustrates his point with a triangle, which places the customer at the highest point, with continuous improvement and total staff participation at the base corners (Figure 4.1). This illustrates the fact that both continuous improvement and the total participation of everyone are key to achieving customers' requests. Total participation is a key point, since most organizations simply focus on a program instead of creating an atmosphere of total participation and therefore do not achieve long-term success.

Because of these factors, Shoji Shiba's Triangle is a great model to illustrate the ethos of an organization at the center of the triangle. How to focus on the customer, total participation of employees, and continuous improvement make up the sides, indicating they are the keys to lasting success. At the heart of the TQM Triangle, we can really see all of the elements forming the values of the organization, resulting in superior performance.

TOYOTA'S TWO PILLARS

Another great example is Toyota's Two Pillars (as seen in Figure 4.2): Continuous Improvement and Respect for People. However, these pillars

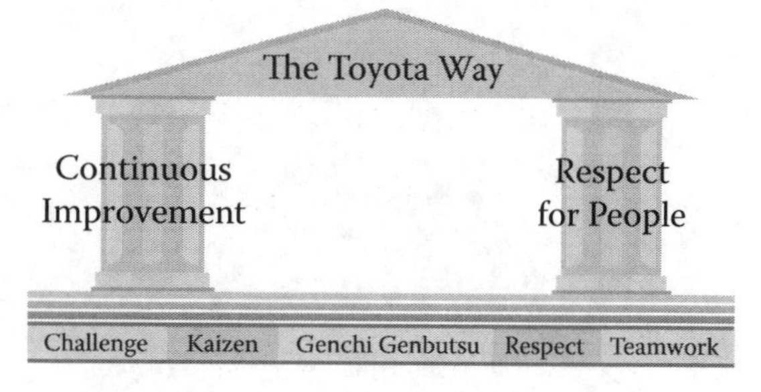

FIGURE 4.2
Toyota's Two Pillars.

represent so much more than most people realize, and not enough literature has been written that focuses on the pillars as philosophical and cultural ideals for Toyota. Toyota views these elements much like Professor Shiba in that, while they are two separate pieces, they are equally as important in order to fulfill the customer's needs. Neither pillar is more important than the other.

Kaizen is a tool that comes from a belief in the idea that everyone, every day, needs a sense of challenge, a sense of purpose about their job, and a connection to their customer. While serving the customer is not a new concept to most people now, not much has been talked about regarding why Kaizen exists as a concept, a mechanism, and overall methodology. What we have discovered is that Kaizen is the tool that brings the concept of challenge to reality. People show what is in their minds, what they are thinking and conceptualizing, in their physical environment.

Kaizen brings everyone together to create ideas and to overcome their challenges. The sense of challenge allows everyone to know that complacency is the root of stagnated progress, and Kaizen is what Toyota uses to connect all people and all layers of the organization together on a daily, weekly, monthly, and yearly basis. You cannot have Kaizen, or Continuous Improvement, without having challenges set out for people to strive to achieve, or without seeking tools and methods to overcome challenges. This is an essential element to Toyota's sustainability and success.

One more important distinction when it comes to how Toyota views these pillars is in how they break each pillar down. Teamwork is a crucial factor to Toyota's success, and Toyota treats it as so much more than simply a tool to get people to solve a single issue (which is how we find most people in western cultures view teamwork). Teamwork is not a term at Toyota; it is an explanation of the organization's structure. There is no one at Toyota who is not on a team. They structure the entire business into modules of teams, from quality teams to career development and customer service teams.

Respect for people is the other element to Toyota's ethos. Toyota explains that respect doesn't begin by giving respect to others; it begins with the realization that before you can give respect to others, you must first have self-respect. Until this is learned, you cannot extend it outward to others. This is considered a Japanese cultural thing, but we believe it is a great lesson that hundreds of other businesses and thousands of people have learned, even if they don't yet identify it as self-respect first, and then respect toward others.

Again, these concepts have not been fully explored by other books, papers, and speakers, so there is a great disconnect between their translations and our understanding. It takes time to explore these ideas in great detail, and every effort should be made to do so because, as managers, you will gain a greater understanding of how these types of visuals impact morale, self-worth, cooperation, and sustainability within your organization.

The well-known author and subject matter expert on Lean, Jeffrey Liker, showed us the rest of the Toyota House is comprised of a number of elements, all of which make the entire system operate like a well-oiled machine. However, it is Toyota's business model that has created the values for the organization, and they have proven it is best to take a holistic approach when creating a values model for your organization.

Jeffrey Liker's visual representation of Toyota's system is not enough to fully describe it, however, as it only highlights the organization's ability to create a reality based on these models. The truth is that Toyota translates this model down to the factory floor and in the office on a minute-by-minute basis. The organization is really living and breathing these tools, and you can physically see it. In any Toyota factory, you will not see the exact same tools and approaches because they don't focus on tools but on principles. These principles evolve at different rates as each generation takes over, and this is true even within a single area of a plant.

Toyota has had more than 100 years to develop its business model, and it continues to change more and more each year. In fact, the Toyota Production System is only one of three models that make up Toyota's Management System. They also have the Toyota Way in Sales and Marketing, as well as the Toyota Engineering and Development System.

These other two models have not been popularized as of yet, but we feel they will be in the future.

WORKPLACE COMMUNICATION

Management almost always views the company in a more positive light than employees do. This same gap exists between higher education institutes and industry; industry sees students as ready for work only 47% of the time, whereas higher-education institutes see their students as 77% ready for industry.* This gap in perception between the two groups shows a need for both sides to be able to understand the opposite perspective and close the gap.

One key area that must be assessed in an organization to close this gap is the communication and actions that happen on a day-to-day basis. Enhancing the ways communication happens within your organization will greatly increase the alignment of goals and people.

While it can be tempting to narrowly define communication as something that takes place verbally or in written communications, everything that takes place is involved in communication. All senses can be involved in interpreting communicated messages; seeing, smelling, and hearing are all part of what make up communication.

> The best use of talent is for people to be able to build an environment that communicates back to them, letting them know how the nonhuman elements of the workplace are working.

Within the work environment, the challenge management faces is how to communicate while also being sensitive to the various messages being conveyed to employees, whether they are intentional, inferred, or unintentional.

> To effectively assess the message a company sends, you need to look at the entire environment of the company, not just superficial aspects of communication, such as slogans.

* http://www.huffingtonpost.com/julian-l-alssid/a-new-gallup-survey-says-_b_4862669.html

As you go through your day, think about all the possible interactions you might have with your employees:

- Do you greet your employees as they come in for the day?
- Do you interact with them before the start of the workday?
- Do you hold morning meetings prior to the start of the shift?
- What kind of message are you sending with your level of engagement at each of these times?

However, there is more to communication than just the interpersonal back-and-forth of day-to-day meetings and checking in. Think for a moment about the visual communications at your facility. What is hanging on the walls in the work area or the break room? Do you have workplace instructions, boards with goals clearly marked, or boards to track daily production?

You probably have at least one poster or sign that states the company's values or purpose statements and of course the required safety and possibly Occupational Safety and Health Administration (OSHA) standards, but is that the only thing you should be communicating? What about why you are in business, who your customers are, or what the benefits of your product or service are? Don't you think those are important for people to know in order to feel more connected to the "why" of their daily job?

There are so many ways in which we can communicate with others, and often, these fall short because we are not doing a good job with the first person who needs to communicate well—ourselves.

Consider how each of these venues for communication affects the content of the communication: personal (1 to 1) verbal communication, written down on paper or a board, or verbal acknowledgement at larger gatherings (think monthly department meetings or yearly addresses).

A simple "thank you" to a factory worker who did a good job would have a different impact if it was stated to them at the end of the day, written on a board in the break room, or mentioned in the president's yearly address.

Remember, venue matters just as much as the content of the message. The best ways to communicate fully, and completely, is to consider the purpose of the message.

There are two fundamental groups to communication. The first relates to the tools to deal with the performance of work elements. The second addresses more interpersonal communications within an organization and is less direct and obvious.

Performance of work elements	Interpersonal communications
Work process procedures	**Teamwork related**
Key performance indicators KPIs Drivers of quality, timeliness, productivity	Team rest area/break room Team boards
Visual examples	**Improvements/solutions**
Ways to communicate without measuring	Sharing ideas
5S for specific items	**Meetings/problem solving**
Placement of tools Storage of materials Picking and ordering Cleaning procedure	Pre-start Breaks Lunch Group How to solve problems as an organization
Closed loop on actions/feedback	**Celebrating activities**
Daily expectations Targets and results	Giving recognition Company sponsored social activities
Tools and equipment designed for easy use	**Management interactions**
Poke-yoke (mistake proofing) Ergonomics	Dialogue with supervisor Visibility of supervisor Visibility of management
How the process and facility has been designed for ease of use	**Displaying values, mission, objectives, and results**
Equipment layout Entrances and exits Lighting Pathways	Displayed and known by all
	Training
	On the job Opportunities for development Memberships
	Safety/welfare of employees
	Facilities and equipment maintenance

FIGURE 4.3
Performance of work elements vs. Interpersonal communications.

Even though these two sets are categorized as two groups they are really working together as a complete communication ecosystem. Figure 4.3 illustrates the high level of interaction that exists between these groups.

Here is a sample of part of Toyota's employee development plan (Figure 4.4). Only a few examples are provided to illustrate, in order to give you some brief insight; however, Toyota shared with us the value of these

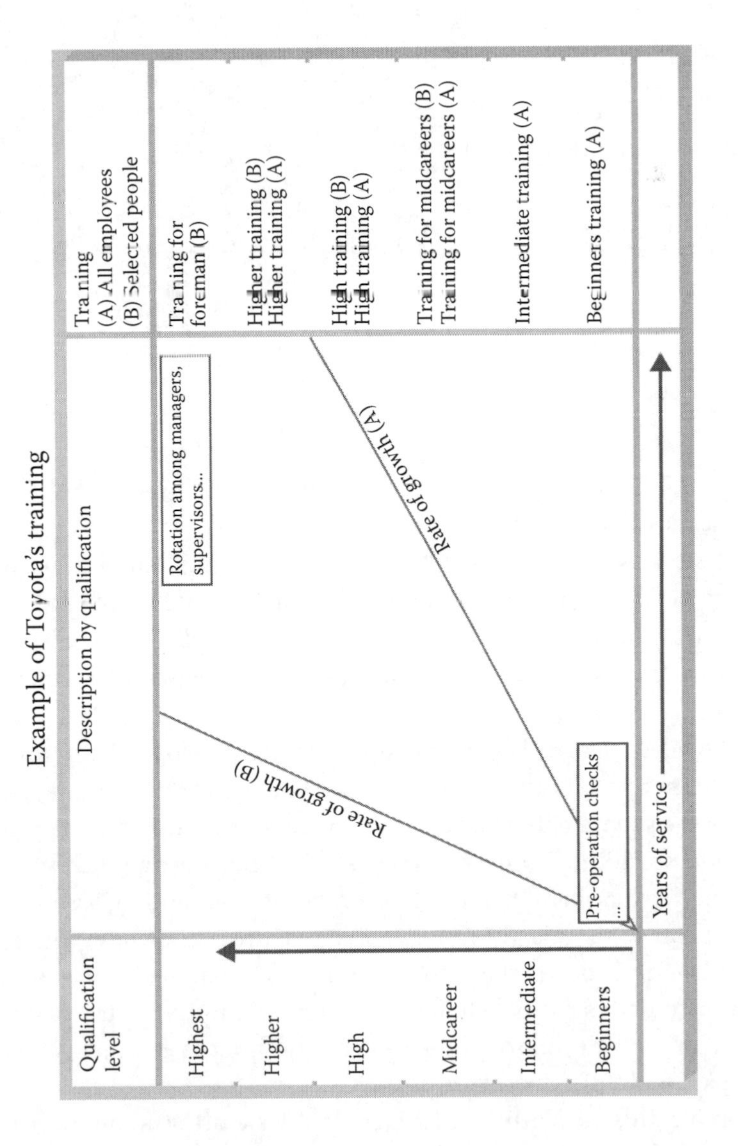

FIGURE 4.4
Example of Toyota's depiction of Figure 4.3.

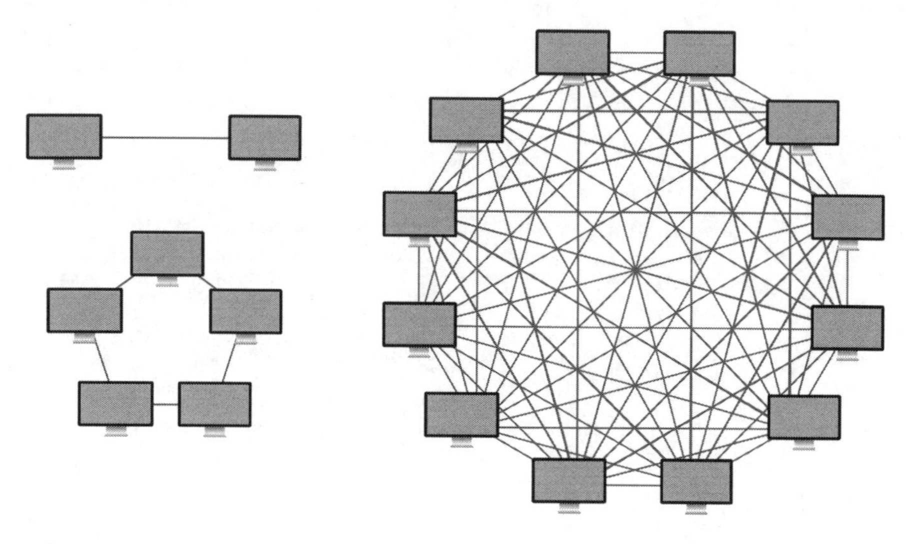

FIGURE 4.5
Metcalfe's law.

groupings, and it is a great way to illustrate to employees the different paths to success for them.

We can never spend too much time communicating; however, we have to be wary of falling into the trap of overcommunicating as well as spending too much time communicating before we know the overall objectives. Objectives allow us to be creative and design systems with the time and resources that we have.

Even though Metcalfe's law* (see Figure 4.5) was developed for hardware, Metcalfe points out that the value, or power, of a network increases in proportion to the square of the number of connections in it. But as managers, you need to spend more time on the design of the system; otherwise, the connections become liabilities and tension points, rather than assets.

Another way to say it is, the more people that are in the know, the larger the benefits. This is one of the reasons blogs and social media have proven so effective in driving social change. By tapping into the ability of everyone to communicate information swiftly and effectively, they generate huge impact with little energy invested.

Combining this with Kurzweil's Law of Accelerating Returns,† which states that technology and evolutionary processes increase in an exponential manner, the ability to communicate is most important. Today not

* http://en.wikipedia.org/wiki/Metcalfe%27s_law

† *The Singularity Is Near* – Ray Kurzweil, Penguin Books, 2005.

finding ways to foster effortless communication will end up constraining the organization and keeping it from performing well.

By making sure to design your organization to consider these models of communication design, you can easily receive communication that is sophisticated yet simple. Communication is essential for success as it makes sure that the organization is able to be agile and respond to the changing demands of the marketplace and needs of the customer.

THE IMPORTANCE OF FRONTLINE SUPERVISORS

Whether it is because of the power of the Internet, allowing people to share information instantaneously, or just the pressures of our modern economy, the workplace is constantly changing. Within this environment, the workforce is now, more than ever, expected to quickly adapt to new knowledge and take action. Frontline supervisors are a key factor in your company's ability to adapt and keep your business agile and effective.

I've seen organizations where frontline supervisors directly oversee up to 85% of employees. As a group on their own, frontline supervisors represent about 80% of a business' management. They are the key contact between upper management and a majority of the workforce. They shape perceptions of your organization, but more than that, because they make up the bulk of your management personnel, they carry out the values you establish as key to your organization. That said, very little business training focuses on these frontline supervisors, as they are often dismissed as just a stepping stone to other roles.

Frontline supervisors are vitally important as they are the first real intersection linking the employee, management, and the customer. They really do set the tone for how the company is perceived by the customer as well as how well the company interacts with its employees. Moreover, the position has a need for unique level of practicality as well as theoretical skills due to this reality.

Studies have found that organizations need a reasonable and scalable mechanism at the smallest level to ensure successful implementations, and that lies with the frontline supervisor. The success or failure of an organization is very often related to the degree of successful change that has occurred at the frontline supervisor level.

Transferring the high-level intent of senior managers to frontline staff is difficult. A methodology for this has been developed within Japanese

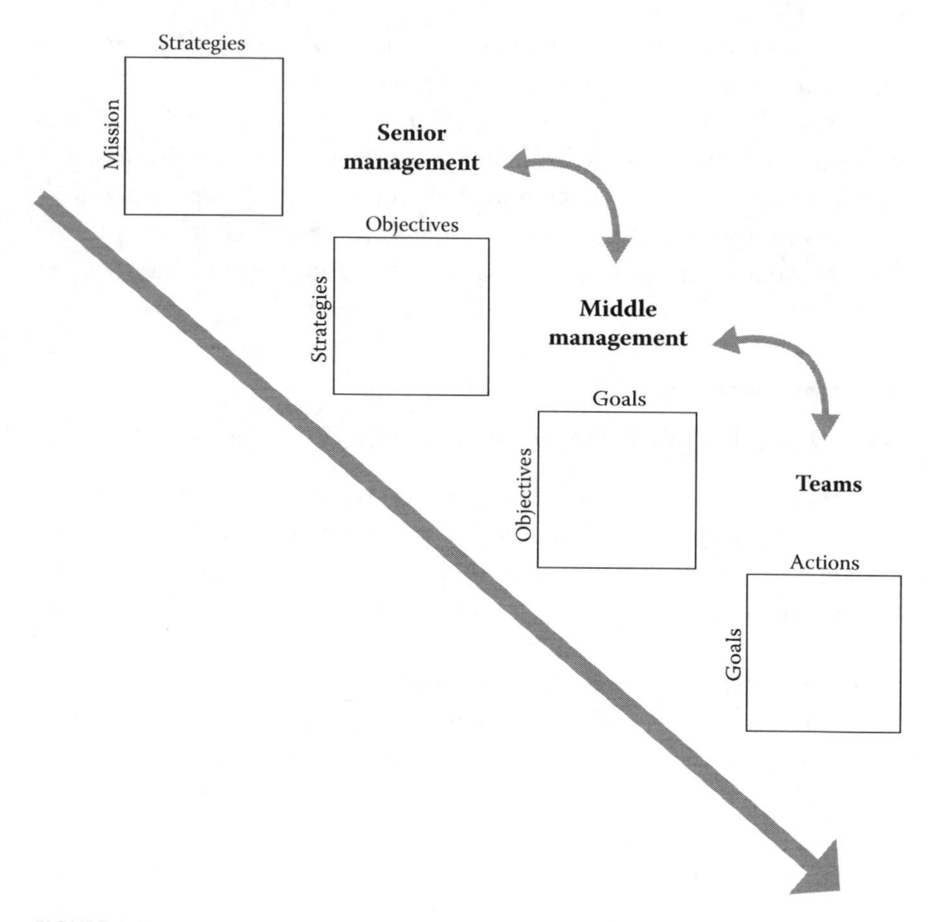

FIGURE 4.6
Hoshin model.

management called Hoshin Kanri. The Hoshin Model (Figure 4.6) summarizes the Japanese method of planning and implementing throughout all layers. The model has been very successful, and it is something that needs to be better developed in Western thinking. We need to develop the what, why, and how part of each person's job as it relates to the plan. We need to connect each person to the overall direction, purpose, and vision to show how one fits in as a contributing member.

According to Professor Yoji Akao, the person who formalized the Japanese method of Policy Deployment known as Hoshin Kanri,* senior management ensure that the strategic goals of the company are driven down to actionable activities at every level of the organization.

* http://en.wikipedia.org/wiki/Hoshin_Kanri

Middle management is responsible to provide senior management with realistic input. Middle management provides senior management with leading indicators to be able to complete the objectives and is responsible to define the "how" of the business systems.

After establishing the benchmarks with middle management, the frontline supervisors, along with their teams, then translate those "hows" into actionable tasks that will accomplish these overall goals. The implementation team then drives a schedule, resources, and actions to reach the goals.

Another way to represent the process is that senior management determines the corporate direction, middle management translates that direction into a plan, and frontline management and their teams carry out the plan in terms of activities. The process is continual, and each level of management makes its respective contribution.

Another model that has been put forward by Daft* illustrated the relative skill set levels of managers and nonmanagers in business. Surprisingly, to most people, there is a considerable need for "human skills" even for nonmanagement roles and frontline employees. This illustrates the importance of "human skills" and communication even from nonmanagers in comparison to managers (Figure 4.7).

This goes against our conventional thinking that "leaders" (managers) are doing the leading, when in fact a lot of leading is occurring at the individual level. Daft shows that there is a lot more leadership, coaching, and mentoring happening than we realize or have been aware of. In Japan, they put a lot more focus on leadership development at the frontline supervisor level, which allows the employees to more directly connect to organizational strategy and leadership.

A great degree of technical skill is expected at the supervisorial level, but the human skills have similar expectations as compared with managers and senior managers. We have even seen some studies showing that supervisors require more leadership/human skills as they directly supervise so many staff on a daily basis. Katz was probably the first to propose this in 1974,† as Figure 4.8 illustrates. While conceptual skills increase at each level of management, the human skill requirement remains the same throughout all management levels and only the technical skills are dimensioned as the conceptual skill needs are increased.

---

* *Management* – Richard L. Daft, Eleventh Edition, South-Western, Cengage Learning, 2014.

† https://hbr.org/1974/09/skills-of-an-effective-administrator

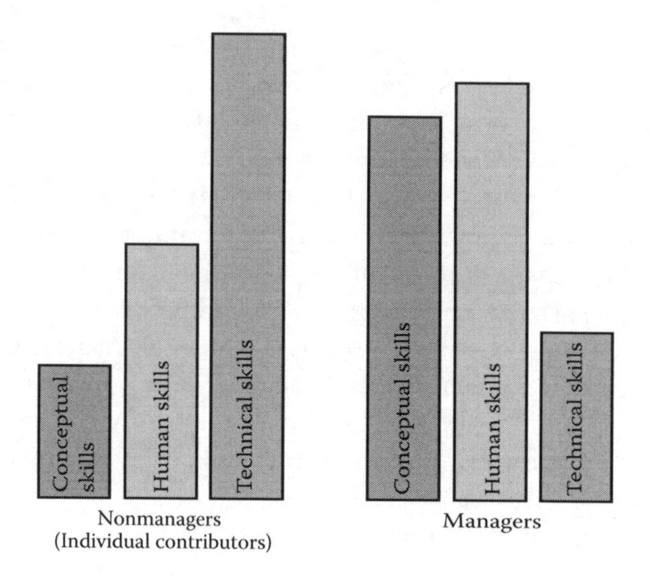

FIGURE 4.7
Example of Daft model explanation.

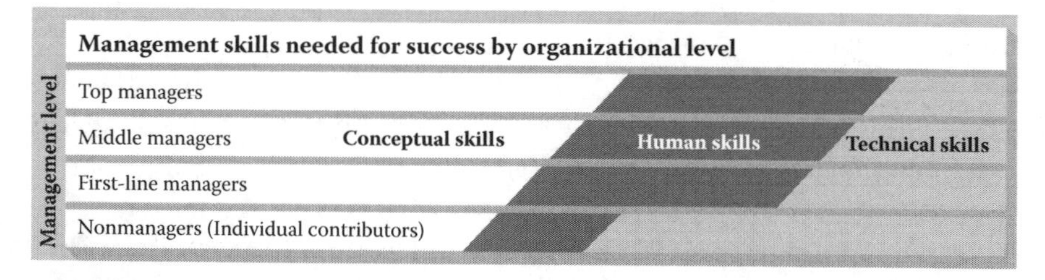

FIGURE 4.8
Relationship of conceptual, human, and technical skills to management levels.

SKILLS TO MANAGEMENT LEVEL

The Daft model illustrates that individuals at different management levels within an organization perform similar activities and use similar types of skills, but the relative mix in time allocation and skills can be drastically different.

The ability of people to fill multiple roles (technical proficiency, people skills, time management), and the ability to interact between employees and corporate objectives and goals is the true role of the frontline supervisor. The general skill set of management functions one needs to balance is illustrated in the following diagram (Figure 4.9).

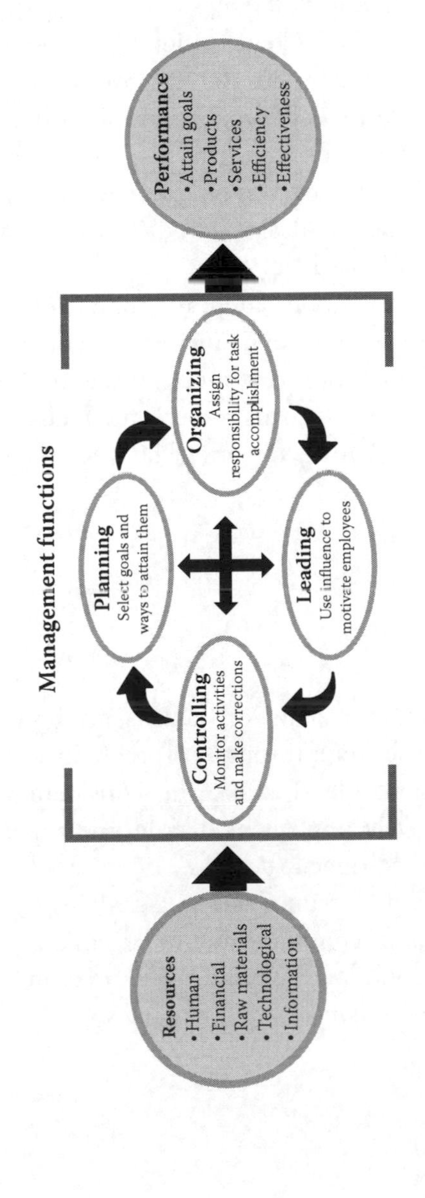

FIGURE 4.9
Leadership model.

These fundamental functions occur at all levels of business and serve as defined core management skills.

These roles are not easy. Previously, we touched upon the complexities of management and how frontline supervisors face a multitude of complex issues, all while having to lead staff daily.

The frontline supervisor is the key position to combine and mold company strategy into daily actions with the employees. How well a frontline supervisor is able to perform is dependent on the degree of development they have received. This position is so critical to the satisfaction of employees and customers. However, this does not mean we can underestimate the need for upper management to design and drive the structures and ethos of success for these frontline supervisors.

Many times, when corporations adopt new practices, this "new way of thinking" gets confused with the dynamics of implementation. The key to implementing any new program is to focus on the interplay between the workers and the corporation. In this respect, the focus should be on the frontline supervisors, as they are the bridge between the corporation and the workers.

CONCLUSION

Because continuous improvements, or Kaizens, are best implemented at the grass-root level rather than through corporate programs, the performance at the departmental level is of prime consideration. In order for programs to be effectively implemented, frontline supervisors are, and need to be treated as, the bridge to success.

Illustrating your company values can be a challenging process, but by using multiple avenues of communication and enlisting the support of your frontline supervisors, you can build an organization where everyone is aligned to your core mission and central values.

5

Management Mindset

The meaning of an active and engaged manager is "a manager that does not allow undesirable situations to remain within the organization." However, most managers have a tendency to not tackle issues and allow adverse situations to remain too long within the workplace. The problem with this is that most often, these situations are things people can deal with and change. They are not technical issues, but interpersonal organizational issues.

In western English cultures, people often talk about the concept of the "Elephant in the Room"—the issue that everyone knows about but no one wants to talk about or deal with (Figure 5.1). This is very often true with managers, as all too often, management does not address the real issues in a workplace, which everyone already knows about. Managers tend to rely on professionalism, when in fact that is not enough. Managers need to be honest about the current state (the reality) of the workplace and work to maintain integrity and hold everyone responsible for achieving the best working conditions and cultural environment.

Taiichi Ohno was probably the best at dealing with these situations and training others to deal with them as well.

One of the key tenets of the Toyota System is the spirit of challenge. This is a key element that demands everyone commit to the idea of breaking through issues and business goals in order to move forward together. This extends into interpersonal aspects of business as well.

Though the idea of "the spirit of challenge" is a good one, oftentimes, managers are still reluctant to place high expectations on their employees. This is for a number of reasons: fear of not being liked, fear of the

FIGURE 5.1
Elephant in the room.

employees not reaching the goals, fear of rejection from peers, and in some cases, even embarrassing their predecessor.

In contrast, Taiichi Ohno is acknowledged as being blunt, focused, and even brutal in confronting situations where his employees were not up to Toyota's standard.

In spite of this reputation, however, Miura's mentor, Yamada-sensei, informed us that Mr. Ohno was a very caring man. He truly wanted people to reach their full potential, but he used his bluntness and confrontational style to ensure the highest success for each individual.

There is a story of Mr. Ohno attending an event at a furniture maker. As he entered the facility, he immediately began directing staff to remove all the shelves. The owner was completely taken aback and asked Mr. Ohno why he was telling his staff to remove shelves.

"Shelves are a source of waste and must be removed," explained Mr. Ohno. "We need to remove the ability to store things in the factory so you can see and achieve the goal of manufacturing to true customer demand."

In the end, the business went from 60 days of inventory to 1–3 days, depending on process and real lead times. If Mr. Ohno had not pointed out the issue, and focused everyone's mindsets on waste elimination, they would have never achieved the goal.

CONFRONTING WASTE

Waste needs to be addressed in order to be a proper steward of resources. This is difficult, but there are a few common approaches: Confront the issue indirectly and gain the support of the people involved informally, or bring it to the forefront and make sure that everything is in the open. Most people try a mix of these two extremes. Taiichi Ohno offered a third approach: Keep everything open, but with a clear purpose and goal that overrides any interpersonal issues.

We have found that it is best to apply each of these techniques as needed to achieve your fundamental goals. Mr. Ohno utilized a multitude of techniques to achieve the organizational goals set to him and to develop people. However, as outsiders to the situations, we only saw the issues that needed to be public, and therefore, it is easy to conclude he had only one approach.

Regardless of what you have read, Mr. Ohno was a change manager. He had the desire and vision to deal with personnel in a manner that was totally customized, while not losing sight of the business and its goals. He built upon respect and confidence in human will and nature. Whichever style you choose, you should never compromise the fundamental goals of changing behavior to create a problem-solving environment.

PROVIDING A CHALLENGING ENVIRONMENT IS ESSENTIAL

Great management has more leadership than management in it. The role of leadership is to work on the process of leading and problem solving in the organization. The role is not to be the problem solver, but instead to instruct, coach, and communicate the techniques others need in order to make decisions and solve problems.

By focusing on teaching others the leadership characteristics as well, you will effectively encourage, challenge, and create a better work environment. Toyota calls this a "culture of solving problems"—a culture that is free to make mistakes in order to learn from them.

The best leaders set a challenge that requires more than one person's efforts. This is key to breaking down the barriers to success. People build common experiences through working together, and this can be far longer lasting, as well as a great experience for all, than a lonely classroom teaching session. In order to set this challenge, leadership needs to find a common goal that unites departments with functions and instills a sense of parallel purpose.

When it comes to problem solving, managers need to challenge individuals to find the real cause of the problem, often referred to as the root cause. Once the real cause is found, managers need to ask what the person believes is the solution to prevent the problem from happening again. To facilitate this, let your team then come up with their own alternatives and provide guidance only when necessary, so that they gain confidence in their skills and you show your confidence in their abilities. This means, as difficult as it may be, you should not provide them the solution. If everything is performing well, iterative challenges can encourage change while matching the business's needs. Managers should constantly test business models, using potential failures to encourage improvement.

SETTING CHALLENGING GOALS

People respond to challenges and constraints, but they also need to receive recognition and celebration in order to have space between their goals and challenges. As a leader, this has to be considered when setting attainable goals. Recognition of achievement thus far, and celebrating that achievement, is just as vital as setting a good challenging yet attainable goal.

When setting target goals, it is also important to remember that there is a distinction between different types of goals: challenging targets, effort targets, and commitment targets. Setting each of these targets should be a part of a manager's repertoire.

Your overall commitment target is a nonnegotiable figure, however, because it is set based upon actual customer requirements, market reality, and compliance issues. This target is undeniable and must be done by all.

An effort target is what can be achieved with a bit of stretch from an individual. It is recognized that this target is demanding beyond what would normally be expected, but with effort and ideas, it is a target that is within reach.

Ohno stated that you have to challenge beyond what is possible. However, remember we are looking at a man who built the largest automotive business in the world, or at least the foundation to 50% of what it is today. So what was his definition of "beyond possible?" We would venture that it changed and matured all the time. When he was first challenged to improve Toyota back in the 1940s, the company was something small and focused, but it was still a challenge.

Toyota started off with one person per machine, and the challenge was to achieve one person per every two machines. This was at one time outside of what seemed possible; however, this challenge created a focus and purpose for operators and managers. It is said that a stamping press changeover, which originally took 16 hours, can now be changed over in less than 3 minutes.

A challenging target pushes people out of their comfort zone. Some would say these targets are impossible to meet, but we must always have a challenging target; otherwise, we can never improve beyond our current capabilities.

New challenges always arise regardless of past efforts. Even Toyota outlined a massive new savings plan to remain competitive after years upon years of improvement and cost reduction. They need to reduce their total resources for vehicle production by 20%. They plan to increase the fuel efficiency of their engines by 25% while increasing power generation by 15%. They also plan to reduce the cost of building any new factories or expansions by 40%. And, by using common parts that require less complex manufacturing, Toyota aims to cut its capital expenditure by 50%. This challenge target is to be completed by 2020.

To come up with an approach to the challenge, there needs to be some foresight and questions asked:

- How can we activate this objective at all layers of the organization?
- What kind of resources do we have available to even consider the implementation of such objectives?
- Can we resequence initiatives to allow us the opportunity to achieve this goal?

Questions like these provide a clearer picture for making challenging goals an achievable reality.

Even managers must have the spirit to consider their own abilities to achieve the challenge target. It is vitally important to the business that managers are able to confront themselves and their own abilities and have the courage to realize their own limitations.

When it comes to addressing your own abilities, there are a few questions to ask yourself:

- Are you willing to engage with the situation? To commit in order to succeed?
- Do you know the operations well enough to provide leadership?
- Are you willing and able to learn new skills to achieve the target goal?

Before you can lead others, and ask them to take ownership, you need to take a look inward. Are you ready to answer the hard questions, really improve your operations, and take action?

> Management requires a purity of intention—you must believe in the work you do and the skills you bring to that task.

When you are in a managing mindset, you must apply the fundamental principles we addressed earlier in this book:

- Problems need not be feared, but embraced. Never blame someone for a company problem. Instead, communicate that the process is at fault, improve the process and test.
- Work as teams in order to accomplish greater things.
- New skills and learning are for all. Management should create structured sharing, where employees learn from each other.
- Organizational transparency is key. It shows that fear is not tolerated. Truth and fixing issues are the goal.
- Visualize information and management. This is difficult but necessary as people will be using all their senses and responding to them.
- Display the organization's values, principles, and expected behaviors so that everyone truly sees and understands what is acceptable and unacceptable.
- The role of management is to continually create a workplace that is open to change and challenges.
- Any issues should be discussed and addressed at the actual place of work (gemba) where value is created.

- Focus on creating models of standards so that if something goes off track, there is a measurable to compare the current state to. Otherwise, you are just focusing on opinions and preferences.
- Create a sense of urgency and work on everyone being more sensitive to seeing and responding. Time is key to minimize an issue. Don't wait—do it now!

As a leader, it is important that you use these principles to build your own form of coaching leadership.

BLENDING KAIZEN AND A MANAGEMENT MINDSET

In Chapter 2, we discussed the meaning of True Kaizen: inspiring everyone to use their own creativity to improve their lives. But what does that mean from a manager's point of view? How do you apply the values we talked about in Chapter 4 with the principles of Kaizen to create success within your organization?

The first step is to begin conducting Kaizen activities. Many consultants will tell you to focus on a "Kaizen Event," typically a weeklong series of activities. The challenge with this model is that it reinforces the notion that improvement should only be done during certain designated periods of time, rather than being a constant process of improvement.

In lieu of a Kaizen Event, consider some of these potential Kaizen activities to begin shaking off the old ways of thinking and begin your own process improvement journey:

Kaizen Suggestion System

As we have addressed previously, it is better for 100 people to take a small step forward than for a single person to take 100 steps. To make this possible, it is extremely important to visualize your Kaizen processes and share the results across departments, in order to establish a genuine Kaizen culture. By drawing out ideas from each employee and applying them, you can create positive synergy, with ideas spring boarding off each other to create more progress than each idea individually.

A Kaizen Suggestion System is a way to measure your organizational culture's progress toward Kaizen. However, there is a fundamental difference in the way the Kaizen Suggestion System was sold to western culture. In fact, it is not truly a suggestion system, as "suggestion" implies you have to submit and gain permission before implementing an idea. This is not what the Japanese actually do. They use the Kaizen Suggestion System to announce the completion of an idea to managers, so that managers can assess and reward employees and communicate the Kaizen ideas of their employees to other managers. This promotes communication and spreads best practices throughout the organization.

> The Kaizen Suggestion System is used to announce the completion of ideas to managers, so managers can assess and reward employees and communicate Kaizen ideas of their employees to other managers.

If you are creating an environment where employees have the freedom to create, to be recognized, and to be rewarded for their valuable ideas, then the ideas system will flourish. If not, the tool is being misused or misunderstood, and therefore, the system will fail. This failure will be because management and the leadership team did not create or support the environment needed for it to work, not because of the lack of ideas from employees.

To help you understand different types of Kaizen Suggestion & Recognition Systems, we have put together a few examples from the companies we have helped and toured.

TOTO

A sanitary wares producer, TOTO has the largest domestic market share of their industry in Japan, and they credit their success to their company-wide implementation of a unique Kaizen Suggestion System. Workers at TOTO are asked to submit a special suggestion form, called the "Instantaneous Problem Solving Form," as soon as they notice anything that could be problematic. The form asks for simple ideas that the worker believes will prevent or solve the issues on the spot (Figure 5.2). TOTO's goal is to have each employee submit 10 suggestion forms every year.

According to a shop leader from TOTO's polishing facility, the number of Instantaneous Problem Solving Forms submitted between March and November of 1 year (a 9-month time frame) was 1654. This is equivalent to an average of 184 per month, or 2205 forms per year. To put this in perspective, they have 150 full-time and 120 part-time employees, which puts their

Kaizen Suggestion Form

Date:

Dept. & Person Name:

Genba Name:

Job Description:

Safety Concerns:

Countermeasures:

Safety Equipment Needed: (Circle)

Gloves Helmet Goggles Shoes
Arm Covers Shin Covers Other: (specify) ______

Special License Needed to Perform Countermeasure?
No Yes Name of License: ______

Name(s) of all Involved:

FIGURE 5.2
Example of a Kaizen suggestion form.

goal (10 suggestions per person per year) at 2700 total suggestions in 1 year. TOTO was meeting 82% of their goal, right from the start of their program.

TOTO's Instantaneous Problem Solving Forms are sorted into distinctive categories for improvement: Quality, Cost, Delivery, Safety, and 5S/Customer Satisfaction. The category that regularly generates the most improvement ideas is the 5S/Customer Satisfaction category, which makes sense because 5S is the fundamental groundwork for all operations, including Quality, Cost, Delivery, and Safety.

The polishing facility draws a variety of visitors for factory tours, ranging from interested personnel from other companies to the local elementary school students, and so cleanliness and safety are a top priority for them. In order to keep providing the greatest experience possible, each visitor is asked to fill out a customer survey form at the conclusion of his or her visit and write down any recommendations or issues he or she identified.

These are taken very seriously by the staff at TOTO and are always utilized for the company's internal problem solving cycles.

Once an idea has been submitted and categorized, it is evaluated by quantifying the effectiveness in terms of how much additional profit can be gained. A "Level 1" ranking is given to ideas that can produce financial benefits of more than 5000 Yen per month. "Level 2" is given to ideas with less than 5000 Yen per month, and "Level 3" is given to ideas that produce no financial results (Figure 5.3). Approximately 10% of worker's ideas toward the Cost category receive an evaluation as either Level 1 or Level 2.

TOTO uses an employee monetary reward system, in which workers can receive a personal financial token for the ideas they generate that are evaluated to be Level 1, yielding more than 5000 Yen per month in benefits to the company.

TOTO also selects Kaizen leaders from each division to participate in a presentation-style meeting every Thursday. At these meetings, each leader has an opportunity to present his or her division's Kaizen suggestion ideas in front of both department and senior managers.

Kaizen leaders select the best Kaizen ideas to be presented, and these suggestions are then called "C-Kaizen." TOTO's distinctive definition of a C-Kaizen is to "Control" and "Change" the current state in order to reach the ideal state by clearly identifying the causes of variation in daily productivity through encouraging employee's "Communication" in order to overcome various "Challenges" to achieve the desired goals (Figure 5.4).

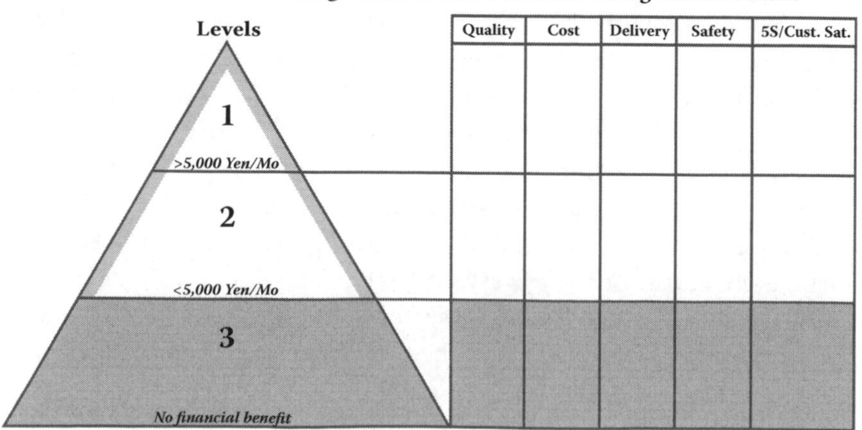

FIGURE 5.3
Example of TOTO's categorization for awards.

C-Kaizen
Control: consider workplace needs **Change:** toward the ideal state **Communicate:** throughout the whole company **Challenge:** breakthrough barriers together

FIGURE 5.4
C-Kaizen at TOTO.

OM Sangyo

Our next example comes from a successful company called OM Sangyo. This company has strived for years to establish an organizational culture that promotes the principles of "action, beauty, and workmanship" for each individual's day-to-day work life. These efforts extend beyond just their full-time employees and out into everyone involved within the company's operations.

Their Kaizen Suggestion System has an additional twist, when compared with TOTO's; along with the individual's unique ideas, their system requires workers to also submit the actual results of the implementation of their ideas. This includes the lessons they learned upon implementing the specific Kaizen strategy they chose as their solution to the problem.

Each report is passed up through the ranks until it is finally submitted to a committee that is designated for spearheading the overall Kaizen Suggestion System within the organization. Members of this committee score each Kaizen implementation report based on their internal criteria, in terms of how quickly an issue was identified, the originality of the solution that was applied, and the level of effort the worker devoted to solving the particular issue.

At OM Sangyo, the number of Kaizen suggestions for 1 year (December to November) was 5292. If you divide that by the number of employees (92 people, broken down into 70 full-time and 22 part-time workers), each worker at OM Sangyo submits an average of approximately 58 Kaizen suggestions each year (Figure 5.5).

In order to keep inspiring workers with the joy of Kaizen and expand their perspectives so that they can better identify potential issues in a proactive manner, the company has in place a technique they call the "TPM Activity Award" (Figure 5.6). This genre of Kaizen focuses on the preventative maintenance of machines. It is focused on increasing the knowledge of the operator

Actual Records of Kaizen Suggestion System (Dec.–Nov.)										
Rating	**TPM Activity Award**	**Issue Identification Award**	**Good Idea Award**	**Effort Award**	**Tin Award**	**Nickel Award**	**Copper Award**	**Silver Award**	**Gold Award**	**TOTAL**
Score	4807	277	187	67	14	2	3	0	0	5292
Entries	90.83%	5.23%	3.53%	1.23%	0%	0.04%	0.06%	0.00%	0.00%	100%

FIGURE 5.5
OM Sangyo's Kaizen Suggestion System.

Criteria	Assessment Standards and Scores Earned				
Identification of Issues:	Identified problems at an individual level	Identified problems at a line/ section level	Identified problems at a department level	Identified problems at an organizational level	
30 pts.	1–6	7–14	15–22	23–30	
Originality:	Even though a solution was not original, the situation is better than before	Results were achieved and met my expectation toward Kaizen	Results were achieved and exceeded my expectation toward Kaizen	New mechanism is put into practice and results achieved by Kaizen	New mechanism is put into practice and results exceeded our expectations
30 pts.	1–6	7–12	13–18	19–24	25–30
Effectiveness:	Visible and easy to understand compared to current state	Visible and easy to understand at a line level	Visible and easy to understand at a department level	Visible and easy to understand at an organizational level	
20 pts.	0–5,000 Yen per month	100,000 Yen per month	300,000 Yen per month	500,000 Yen per month	
Effort:	Average	Relatively High	Much higher than average	Higher than expectation	Much higher than expectation
20 pts.	1–4	5–8	9–12	13–16	17–20

FIGURE 5.6
Example of OM Sangyo's TPM Activity Award criteria.

because many maintenance tasks that need to be performed are neither mechanical nor engineering, and therefore, they should be known by the person who is working on the machine. Small rewards are given as incentives, but the key to this system is recognition and formalizing the process to bring true integrity to the system (Figure 5.7). This TPM Award system allows workers to actively submit reports describing solutions that they have implemented for preventing specific issues from happening, even when their solutions may have been as simple as cleaning, clearing items, repairing, or labeling items.

While the effectiveness of such tasks that prevent issues from happening may be difficult to quantify in terms of overall company-wide goals, it is often small solutions that prevent serious issues from arising. In fact, the TPM Activity Award accounts for 91% of all the awards given to the workers each year at OM Sangyo, and the company continues to appreciate and promote these small, simple solutions. In doing so, they effectively eliminate much bigger issues in the long run.

A team leader from OM Sangyo emphasized that even though this cycle is called the "Kaizen Suggestion System," it doesn't stop with mere suggestions. It is a process of confirming that Kaizen ideas have been actually implemented as planned.

Sinko Air Conditioning

Next, we want to introduce you to Sinko Air Conditioning Industries, who have significantly expanded upon the idea of the Kaizen Newspaper.

Similar to TOTO and OM Sangyo, Sinko assesses their internal Kaizen Suggestion System ideas based on the level of financial benefit each Kaizen implementation has effectively brought to the company (Figure 5.8). However, only the Kaizen reports related to Safety issues are scored in a qualitative financial manner. The scores are then published in a monthly internal newspaper, called "The Kaizen News."

Total scores are calculated every 6 months and individual workers are ranked from 1st to 10th place based on the scores they have achieved. The workers are listed along with their respective department. The top 10 individuals and the highest ranked department are later awarded with prizes in front of everyone at a morning assembly (Figure 5.9).

The company also has a second achievement recognition system called the "Kaizen Grand Prix," which highlights the quality of Kaizen ideas from

Monetary Award Criteria for Kaizen Suggestion										
Rating	**TPM Activity Award**	**Issue Identification Award**	**Best Idea Award**	**Effort Award**	**Tin Award**	**Nickel Award**	**Copper Award**	**Silver Award**	**Gold Award**
Score	4–9	10–14	15–29	30–39	40–54	55–69	70–79	80–89	90–100
Prize	100 Yen	300 Yen	500 Yen	1,000 Yen	5,000 Yen	9,000 Yen	14,000 Yen	20,000 Yen	30,000 Yen

FIGURE 5.7
Example of OM Sangyo Monetary Award criteria.

Example of Sinko's savings evaluation criteria			
Evaluation	**Designing dept./ operation dept.**	**Other depts.**	**Kaizen related to safety**
Grade 1	More than 10,000,000 Yen	More than 5,000,000 Yen	Prevented accidents that can cause the death of employees
Grade 2	More than 5,000,000 Yen	More than 2,500,000 Yen	Prevented accidents that can have serious prolonged side effects
Grade 3	More than 2,500,000 Yen	More than 1,000,000 Yen	Prevented accidents that can suspend business
Grade 4	More than 2,500,000 Yen	More than 500,000 Yen	Prevented accidents that can injure employees
Grade 5	Over 100,000 Yen		Prevented accidents that can involve workers falling
Good idea award	Over 50,000 Yen		Prevented worker's injuries such as scratches or cuts
Good spirit award	Less than 50,000 Yen		Great ideas, but not implementable
Participation award	Not implemented/ repeated ideas		

FIGURE 5.8
Sinko's evaluation based on cost savings.

Monetary Awards given in half year period at Sinko						
Top 10 employee rankings	10-6	5	4	3	2	1
Monetary Award amount	2,000 Yen	3,000 Yen	5,000 Yen	10,000 Yen	20,000 Yen	30,000 Yen

FIGURE 5.9
Award money given in a half-year period at Sinko.

workers. It helps them all strive toward a higher level and boosts overall morale (Figure 5.10).

Kaizen Grand Prix	
Types of awards	**Award description**
Newcomer award	Applicable to workers who have been with the company less than 2 years
Karakuri award of excellence	Best example of implementing Karakuri devices in machine Kaizen
Award of excellence	Excellent example
Grand prix award	Most excellent example

FIGURE 5.10
Sinko's Kaizen Grand Prix.

For the Kaizen Grand Prix, management evaluates over 170 workers for their total number of Kaizen reports, the quality of their reports, and overall "good spirit" of the employee. Their record so far in total number of Kaizen reports submitted in a year is 3359 reports, which averages to about 19 reports per person that year. Most reports have received Sinko's Good Spirit Award, which goes to show how far Sinko Air Conditioning has gone to create a great workplace that engages their workforce.

At your facility, think beyond how you can solicit ideas and begin to think about how those ideas are received. How can you welcome and promote ideas in such a way that encourages more engagement and generates excitement at your facility?

Kaizen Activity Visualization Board

Another Kaizen tool that can be utilized to great benefit is the Kaizen Activity Visualization Board. Posting Kaizen scores and ideas on a public bulletin board, where everyone can see the suggestions, will increase workers' awareness toward Kaizen. These should include not just the ideas but also the before and after (to highlight results), as well as the idea's effectiveness of reducing costs.

Kaizen implementation reports are also important to promote cross-departmental education and foster the sharing of great, practical ideas. Workers will be able to learn from each other and become more skilled at identifying potential issues through examining their colleague's ideas. This display will promote a Kaizen culture that will continue to spread outside of specific departments. Ideas can lead to newer ideas and can yield much greater results through applying the next level of Kaizen strategies.

Department-Based Kaizen Boards

Yamaguchi Seiki Kogyo

Another way to promote a Kaizen culture is to require one Kaizen report submitted each week by each department. An automobile plastic die manufacturing company, Yamaguchi Seiki, has implemented such a system in their facility. The result of the pressure of displaying their weekly Kaizen idea on a large public board has stimulated the people within each department to strive for higher understanding of Kaizen. In essence, they have effectively harnessed the basic competitive nature of humans to benefit the company and each person's day-to-day job.

As illustrated in Figure 5.11, all the departments (including design, machining, finishing, and all administrative divisions and managerial roles) are listed to present their Kaizen ideas to everyone once a week. At an early stage of this weekly Kaizen Visualization Board implementation, there were some departments that diligently complied and others that were not following the requirement at all. Since Yamaguchi Seiki's board stands at 4 meters wide and 2 meters high (or roughly 13 feet by 7 feet) and is installed in front of the facility where the daily morning meeting

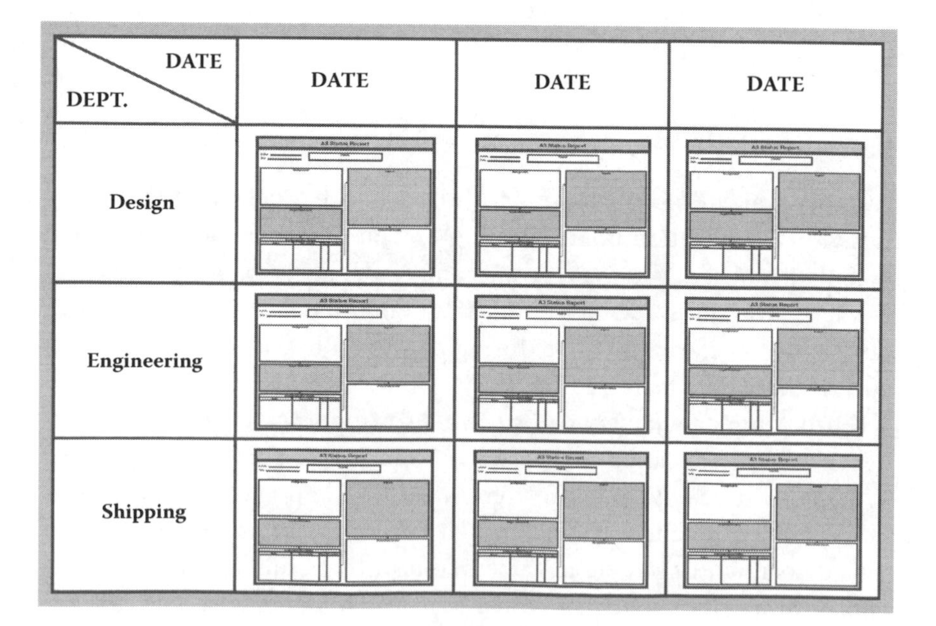

FIGURE 5.11
Example of Yamaguchi Seiki Kogyo's weekly Kaizen visualization board.

takes place, it is obvious to everyone which departments are following the rules and which are not, and this helps to keep people accountable.

One important lesson that Yamaguchi Seiki has learned in implementing this visualization board is that the first thing to do when eliminating waste (Muda) is to instill a strong desire in everyone's mind to want to remove waste.

As people's mindsets change, their actions become faster and more responsive, and as people's actions become habitual, they learn to execute Kaizen in a competitive manner with each other, each and every week.

For example, the design area is divided into four teams, and each team is assigned a particular week within the month when they spend most of

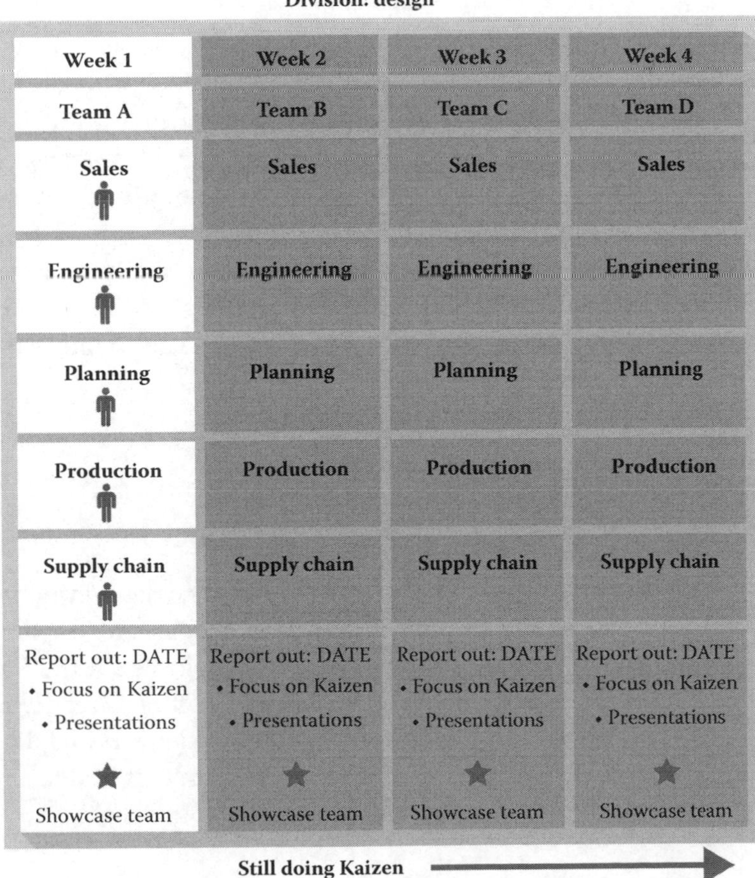

FIGURE 5.12
Example of team creation at Yamaguchi Seiki Kogyo.

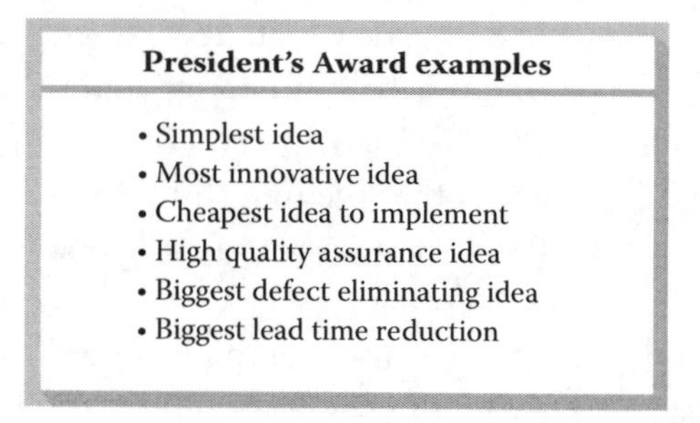
President's Award examples

- Simplest idea
- Most innovative idea
- Cheapest idea to implement
- High quality assurance idea
- Biggest defect eliminating idea
- Biggest lead time reduction

FIGURE 5.13
President's Award at Yamaguchi Seiki criteria.

their energy concentrating on implementing Kaizen ideas (Figure 5.12). This ensures that the division stays actively engaged in reporting out their Kaizen ideas.

At the end of each month, a "President's Award" is chosen for the area whose idea was deemed the most effective and breaks through the old ways of thinking the most. The president of Yamaguchi Seiki makes the decision himself and presents the chosen department with a small gift and award at the beginning of the next month, celebrating the achievement with everyone within the organization (Figure 5.13).

The secret behind this company's weekly Kaizen visualization mechanism is that it makes sure that each division executes a Kaizen idea every week and encourages workers to constantly challenge one another to create a true bottom–up Kaizen culture.

CONCLUSION

With the right management mindset, you can begin changing your workplace toward a Kaizen culture. The ideas in this chapter are just some of the tools and techniques to inspire and engage employees in the Kaizen process. Having examined the why, or the mental shift required for management success, we will shift gears in the following chapters and dig into the physical changes you can make to your workplace that lead to a more efficient operation. And those changes start at the front door.

6

How to Assess Shop Floor Morale

Roll up your sleeves and join us as we show you how to take the principles outlined in the first half of this book and apply them at your facility. The first step is to assess the current state of your operations. Where should you do that? On the shop floor, of course!

When you go to the shop floor to begin determining how to apply Kaizen in your organization, the first question that should come to mind is, "How should I assess the current state of the workplace?" To answer this question, Miura takes the companies he coaches on factory visits to different industries and has them learn how to read and assess a plant by using Personal Education Center for Kaizen's (PEC's) unique factory assessment sheet. (PEC is where Miura works. Enna, through Jun Nakamuro's work, has been officially endorsed by PEC's founder to carry on Taiichi Ohno's only training program.)

This assessment allows people to benchmark their own factory operations against those of the host facilities they visited. If these trainees—as they are designated when working with PEC—score their own operations higher than the host factory on certain criteria, they will give the host company feedback and share their Kaizen ideas with them, in order to help improve the host company. If they score themselves lower than the host company, they can receive the same type of Kaizen suggestions and feedback from other trainees, as well as the host factory. The key is that every trainee learns how to benchmark themselves against others' operations and everyone can improve together through Kaizen idea sharing and feedback.

The criteria that are used to assess a factory are as follows (Figure 6.1):

Morale: The level of commitment of employees to their actual work
Stagnation waste: The amount of work-in-process and inventory stored across the facility
Motion waste: How each person moves
Transportation waste: How each person performs a task
Management systems: Criteria to check Kaizen activities, production management, equipment maintenance, quality assurance, and safety

PEC's training program is a total of 6 months long, and the trainees attend sessions 2 days each month, at a different location each time (they rotate through the trainees' companies for the events). This allows them to focus solely on Kaizen implementation practice, learning to see wastes, and putting their knowledge and skills to the test, without having to devote too much time away from their normal jobs.

The training program begins with morale training to help the trainees realize that change starts with them. Next, they begin working on rapid change, complete assignments at their own organizations between

Assessment criteria		Assessment points
Morale	Morale	Greeting, dress code
Stagnation waste	How items are placed	Sort, set-in-order, store, fridge
Motion waste	How workers move	Motion waste, work speed
	Multiskilled workers	One person production, multimachine production, multiprocess operations
Transportation waste	Transportation	Just-in-time, forklifts, carts, WIP
Management system	Kaizen activities	Policy deployment, sharing of Kaizen results
	Production management	Production management board, shipping management board, customer promise date
	Equipment	Changeovers, maintenance, poka-yoke, jigs
	Quality management	Quality inspection tools, defect handling, built-in quality in the production line
	Safety	Work environment, lighting, protective gear

FIGURE 6.1
PEC factory assessment example.

training events, and then rotate together through each trainee's company (one per month) to work together on solving issues.

For each day of the program when they all are together and learning in a workshop setting, they participate in report outs to share what they have achieved with their fellow trainees, PEC staff, and the business management of the host company. They celebrate their achievements and close out the day with an awards and recognition ceremony to show the trainees, by example, how good recognition is achieved, which is something they bring back to their own organization and share.

This shop floor training helps the trainees understand how to yield results by executing their own Kaizen ideas right away. This quick execution of ideas means that, when they leave the host company, they have made a substantial impact to the company in terms of Kaizen improvements. They also come away with a greater personal understanding of how to become trainers within their own organizations. The host companies also benefit by receiving Factory Assessments, filled out by trainees, showing where additional Kaizen improvements can be made, based on input from people external to their day-to-day business.

This benchmarking process allows each company to keep striving to level up their Kaizen implementation to achieve much higher goals than they thought were possible.

MORALE

As we discussed in Chapter 3, nurturing a spirit of commitment toward work within each individual is essential to increasing product quality and productivity. However, this level of daily commitment cannot be calculated and reported out in the final financial reports of your organization. So how are you supposed to measure the commitment of each person within your company?

PEC uses a method of measuring worker's commitment by assessing their morale and defines a workplace with a high level of morale as follows*:

- People greet each other with a loud voice and good choice of words.
- People move briskly and with a purpose.
- People have a purpose and goal.

* Hitoshi Yamada, *Hitome de Wakaru, Sugu ni Ikaseru, Kiso kara Wakaru Kaizen Leader Yosei Koza*, Nikkan Kogyo Shinbunsha, 2008.

A healthy greeting in the morning can break the ice and instantaneously change the atmosphere of a workplace toward a more positive state. Work cannot by completed by individual effort alone but must take place through the collective effort of teamwork. A warm greeting is so simple that anyone can perform it right away with no training needed, and it serves as a way to promote excellent human relationships in your workplace.

Your colleagues and employees may already greet each other, but in order to make the most impact, pay more attention to how it takes place and assess if their greeting makes you feel refreshed and is being done in a way that establishes mutual respect.

This type of warm respectful greeting was a characteristic of Sam Walton and was an important value he promoted at Walmart. There, it is a standard to have a Walmart employee greet customers at the entrance and within 10 feet of entering the store. This is now a rediscovered practice by many stores that have been under pressure during the most recent economic downturn.

In addition to greetings, team members must be able to clearly answer customers' questions, while also being able to clearly provide explanations. This is important because it shows a transparency of information and trust in employees to take care of the customer, whether internal or external to the company.

For example, in the instance of discussing a topic with the operations manager of a company that Miura works with, the lead person was less than forthcoming with his answers. What was the underlying message? Was it insecurity on the part of the lead person, or mistrust of the manager? Suddenly, Miura's focus was pulled from the operations and instead was focused on the unusual interaction between employees. Such behavior should be an immediate concern for a manager.

In addition to how people act toward one another, the brisk movement of people in an organization is directly tied to the level of productivity that is achieved by each person. Let's say a person needs to walk five steps to reach a particular part or tool. If said person reaches the part in 3 seconds instead of 6 seconds, that means their productivity is increased by 200%.

People typically walk at an average speed of 3 mph, but if we instead walk at 4 mph, we can enhance productivity by 33%. Active movement also increases blood flow to the brain, stimulating a more positive attitude. Brisk and quick movement is the foundation for improving productivity and is an important part of PEC's training program.

Another area to assess is dress code. Uniforms that promote cleanliness, safety, and workability in your environment are highly recommended. By uniforms, we don't mean everyone at every job within the company has the exact same outfit, but that all are similar in look and feel. This is important because it helps people feel a sense of identity within the organization and show pride in their workplace.

The Japanese find that preparing for the day, getting themselves in the right mindset, and fostering a sense of teamwork can be more easily achieved in a set, prepared appearance each day. In putting on a uniform, keeping it clean and tidy, and being proud of the workplace you and your coworkers have built, you will project a sense of pride outwardly. Your coworkers and customers will feel this and ultimately respond back in a positive manner. A sense of identity, pride in place of employment, and care in personal appearance are as much a reflection of the organization's values as of the attitudes of the team members.

Morale can be a tricky concept to get a handle on, but enthusiastic greetings, active movement, and dress code can be good indicators of morale level. Next, we will go into greater detail by providing a few examples of how these principles of morale helped improve real companies.

ENTHUSIASTICALLY GREETING THE DAY

An electric connector company, Hirose Electric, has seen their profits rise by more than 20% over the last few years. They attribute this to a rise in overall company morale, which began after one of their factory section managers, Mr. Yoshihiro Gunji, attended one of PEC's courses on morale training. During this course Mr. Gunji screamed at the top of his lungs (part of the training that happens to teach people about breaking through mental barriers) and learned to move briskly with a purpose.

When he returned to his facility, Mr. Gunji decided to make changes in the way things were done, starting with himself. He began greeting each of his 27 employees with a hearty "good morning!" in a much more energetic way than usual. After the initial confusion over his change in style, his employees began to feel extremely refreshed and energized by the greeting.

Everyone realized how much a morning greeting could dramatically brighten up the atmosphere and break down communication barriers between colleagues. Before long, everyone in the department was enthusiastically greeting each other every day. After a while, they noticed that, as the volume of their greetings had increased, so had everyone's response times to their daily work. Mr. Gunji was delighted to find that an energetic greeting could substantially enhance the morale of his employees, and even impact productivity.

GREETING VISITORS

An enthusiastic greeting not only affects your employees, but it can also have an effect on your customers and visitors. A wood furniture company, Kashiwa, welcomes a number of factory tour visitors, including current customers, onto their production floor every month. The part of the pathway that runs alongside the first process quickly became known as "Greeting Avenue," as each machine operator pauses in his or her work and energetically greets them with a "welcome to our factory!" After greeting the visitors, workers then go right back and continue their work.

When visitors from other companies and countries visit, the company names and country flags are displayed across the shop floor, in order to make visitors feel at home. This is also Kashiwa's way of showing respect to their visitors.

These efforts don't cost Kashiwa much in terms of labor or effort; the amount of time each worker spends greeting visitors is only around 5 seconds. However, those 5 seconds make a huge impact on the visitors, and the display helps them understand that the furniture they are about to purchase is being manufactured by people who are happy and genuinely enjoy their work. It is clear to every visitor that the people who work at Kashiwa have a customer-first mindset, which creates a deeper relationship between customers and the company.

MOVING IN THE MORNING

The employees at the headquarters and main production facility of Sinko Air Conditioning, in Okayama, Japan, begin their day with a group exercise. Once everyone is warmed up, they go through a routine that

FIGURE 6.2
Walking the Ganba-Road.

reminds them of what it feels like to walk briskly with a purpose inside the plant.

This routine centers on a white line, approximately 150 feet long, drawn on the floor. They call it the "Ganba-Road" (This is a play on the word "ganbaro" which means "do your best."). Each worker walks along the white line, with a goal of reaching the end within 30 seconds. If a worker can complete the walk within 30 seconds, it means they are moving at almost 4 miles per hour—a very brisk pace (Figure 6.2).

The assistant to the department head, Mr. Tomonari Ando, tells factory visitors that, "Going through this exercise every morning allows workers to internalize the appropriate pace at which they should be moving, to achieve enough briskness." In other words, they physically set the standard within themselves each and every morning.

MORALE AND PHYSICAL WORKPLACE CONDITIONS

Another way to see how the physical world represents and affects the morale of an organization is by looking at the overall physical appearance of a place. First impressions are easy to make and difficult to undo, and it is far easier to reinforce a good impression than overcome a poor one, so it is vitally important to an organization that their physical space is well received.

Organizations whose goal is entertainment, such as Disneyland, California Adventure, and even Las Vegas, are probably the ultimate leaders in creating an environment that allows people to enjoy and achieve extraordinary things. Even Cirque du Soleil brings such a high level of performance and atmosphere that we forget our daily lives for a while and simply enjoy the experience, from the moment we first arrive.

Such organizations are assessing their operations by the minute to make sure they are constantly maintaining the right environment for the respective themes. In many places, these factors aren't happening because of intentional focus. However, if we put intent into what we are as a collective group, people will naturally be attracted to it because it will be as enjoyable to the customer as it is to the employee.

Managers at all levels and in all forms of organizations, from churches to amusement parks, bus stations to factories, should be fully cognizant of their organization's impact on human behavior and attitude.

The appearance of your facility, including the grounds, outside buildings, inside offices, and interior facilities, is a reflection of your values. Look at the impact of repurposing buildings in older towns. Many towns are now focusing on renewing old areas of town to attract new commerce opportunities and lift their economy. Old factories are being converted into apartments and condominiums, office space, and even fashion-forward new stores.

What makes the difference is not the fact that the buildings are new; the purpose and nature of the people within the buildings are what have changed. They are improving something that is old and full of character, and they connect themselves to this cause and invite others to join in. The buildings remain the same; it is the people that make all the difference in environment.

Similar factories may make similar products, but odds are they will have wildly different environments. The details in each workplace convey the inherent values of the organization, and these in turn convey the true ethos to others. Think about the previous examples and how an overall business extends beyond the physical building to include people's attitudes, the sounds, kinds of safety equipment, and even what materials are used to make small changes. Is the business using simple things to create trials before committing to large costs and designs?

So where should you begin assessing the appearance and environmental impact of your facility? Easy—the front door!

BUSINESS ENTRANCES

The first contact point everyone has with your facility is the entrance. If you see the entrance is in good shape, with the lawns and landscape well maintained, it is already sending a positive message. In fact, one of Collin's best experiences was at a hydraulic motor manufacturer in Greeneville, Tennessee, called Parker Hannifin. The biggest impression of value to Collin, as a guest, was the first contact with a security guard. The guard knew that the visit was scheduled for that day, so the guard offered Collin a bottle of water as soon as he arrived, had made the badges for the day, and knew the nature and purpose of the meeting and all who were involved. This all took place while the entrance was kept well maintained, as if no one was visiting that day. Even before talking to any management, this forms a great first impression of any business (Figure 6.3).

ADDITIONAL EFFORTS

Another way you can improve the image of your facility is to just put a little customization into the event. Putting up flags, additional signage for the event, or even special annual banners and awards will go a long way.

You can fly the flag of your country, your company flag, a regional or state flag, or even the country flag of the visiting customers of the day. Figure 6.4a and b shows examples of flags being flown inside of factories for tours as well. This is a special recognition of the visitors and is something everyone will notice and appreciate.

PARKING LOTS

Even your parking lot gives off signals and has an effect on people's perceptions of your site, not only in cleanliness but also in what is important

FIGURE 6.3
(a) Yamaguchi Seiki entrance. (b) Hida Sangyo entrance and greeting.

(a)

(b)

FIGURE 6.4
(a) Using flags with welcome signs. (b) Tour guest welcomed by his country's flag.

to the company. All too often, you see parking lots with specially designated spaces for certain individuals, which is fine to have, but what purpose does it serve? Are there special spaces reserved for executives close to the entrance door? How about the employee of the month, expectant mothers, handicapped spots, or emergency vehicle parking? All of these send a clear message about the company's culture and what is a priority outside of normal production.

INTRODUCTION TO THE COMPANY

A great introduction to any company is an information board, situated in a place that is best suited for both employees and visitors. On our last visit to TOTO in Fukuoka, they had embraced this on a massive scale and created a board showing operational metrics, safety, quality, productivity, new product development, and a variety of other key measures. They placed it where all employees and visitors pass by to convey what is important to the organization (Figure 6.5). Even at your local mall, you will find kiosks that show directions, a business directory, and other important information necessary for customers. Why not do the same at your company?

FIGURE 6.5
TOTO's entrance hallway in Fukuoka.

FIGURE 6.6
Personalized welcome signs.

WELCOME SIGNS

The best businesses are able to communicate visitor information to staff so that the organization can appropriately greet expected visitors. It is a nice gesture and there will never be return on investment (ROI) from this specifically, but it is our experience that it provides a great impression to the visitor or customer. This doesn't mean you have to install a special electronic board or anything special, and we have no doubt you will find that your employees will put their own creative energy into developing a welcome board (Figure 6.6).

Managers and supervisors can also use this to inform their employees of visits, as it will be obvious to them as they enter the facility. This is another great example of effective communication.

DEPARTMENT BOARDS

Department boards serve a number of purposes, including showing employees and visitors vital information such as production goals, quality reality, safety issues, product mix, and shipment data. The mission statement and

FIGURE 6.7
Performance recognition board.

purpose should also be there as a daily reminder because exposure to this information reinforces the value of the organization and the daily reality of production (Figure 6.7).

Postings can also include information such as employee of the month, special recognition and awards received by team members, special interest features, and anything else that reinforces a focus on a caring organization.

PURPOSELY ARRANGED OPERATIONS

The inside of the facility must be kept clean and free from any obstructions. Your checklist of items will vary slightly for each workplace and organization, but in general here are some basics (Figure 6.8):

- Lighting is adequate, no burned-out bulbs and appropriate lighting to see work
- Minimal noise, dust, smells, vibrations, fluids etc.
- Clearly marked aisles, free of obstructions
- Clearly marked staging areas for inventory, parts, jigs, and equipment
- No excess items, materials, or parts
- Ergonomic use of vertical space throughout
- Tools and fixtures in their respective marked locations

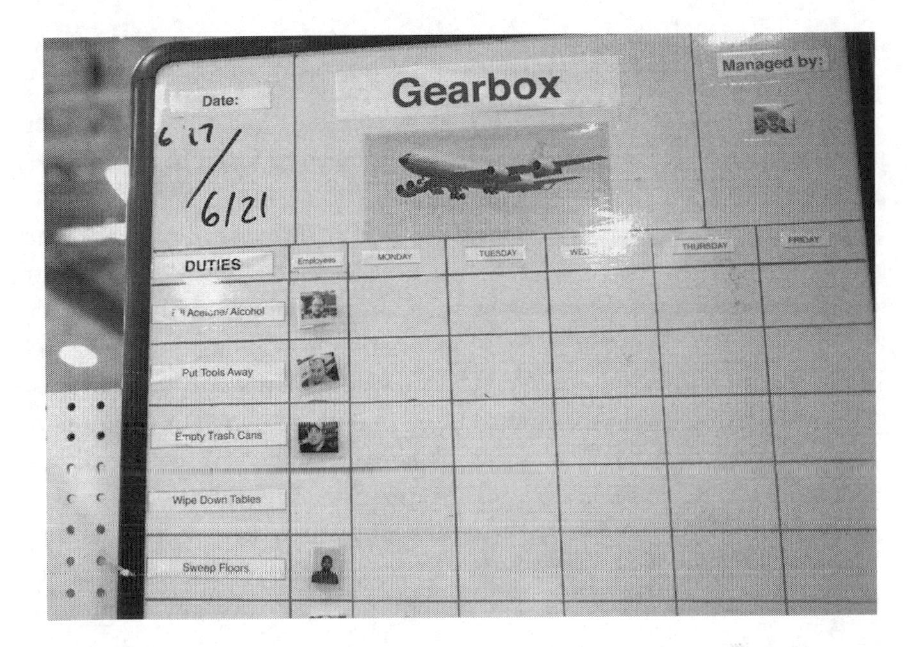

FIGURE 6.8
Example of housekeeping duties.

An organization's ability to maintain a clean work environment is a testament to how well production and daily demands are managed. If the facility is clean at 10 a.m. on a Thursday, it shows the operation is well managed. Gallup has stated for years that the attractiveness of a workplace is linked to both engagement and productivity. If the plant is kept meticulously clean, where dust bins, mops, scrap, and supplies are clearly available and controlled, this provides the best message to employees and customers about the value of cleanliness.

One business, White Water West Industries, went so far as to use their facilities to show customers how they control and manage production, so cleanliness and safety are of vital importance. Their customers were so impressed with the process of flow and management that the operations floor became their competitive advantage to making a sale. Now that is connecting production employees directly to customers! Think how much more fulfilling this is to their welders and painters—showing the direct connection between their hard work and the customer's appreciation.

As we discussed earlier with Mr. Gunji of Hirose Electric, personally greeting everyone each morning is vitally important. People need to be greeted and feel that they are important. Managers and supervisors can help to foster each person's sense of worth at work by an upbeat and sincere greeting.

Another relevant tool, which brings people together and helps foster interpersonal relationships, is the stand-up meeting to go over the day's plans and other relevant information.

At an insurance company Collin coached in Brownsville, Texas, the general manager was stressed out over the fact that some stores were far more productive than others. When Collin first arrived at the business, the manager was sitting at his desk reading reports about the stores and simply complaining. So Collin suggested that perhaps selling insurance appeared very easy to him sitting at his own desk, looking at the numbers all lined up, but by actually going and seeing he might find that the difference in numbers and reality of each location would reveal itself.

Upon going out and touring the different stores, it was obvious that the employees were not connected to the organizational objectives at all. Some did not even know who the general manager was, and believed it was Collin.

Moreover, the general manager was not touring each store enough and conveying the purpose and challenges of the organization. Eventually, Collin told him not to share the numbers the employees needed to meet, but the challenges. Some stores were better laid out, some had more traffic, some had faster systems, etc.

What the general manager then did was make the challenges of each store clear and apparent to all of the stores. Taking that type of information to the staff, he toured each store and challenged them to overcome the natural barriers to their store. They all rose to the occasion. Each store had natural barriers to flow and they found unique ways to overcome them. Their disadvantages became talking points for improvement, and once they began to overcome their barriers, they began sharing to achieve even better measurements.

In fact, the store with the lowest volume of customer demand innovated to get more business from each customer. The higher customer flow locations could not focus on this same outcome, as they were focusing on reducing waiting time.

This is the kind of knowledge and communication that numbers and summary reports do not provide. Moreover, because the general manager exposed and shared the truth of the situation, the employees were able to use their own creative ideas to find success and overcome barriers. Within the following 6 months, all stores were higher producers and had less staff turnover.

In another instance, a very famous person in Lean circles, Paul Akers, moved his entire office staff onto the shop floor and removed all meeting rooms. This meant that everyone now had access to everyone else and had visibility over the entire process. At his company, FastCap, problem solving and waste elimination are more important than production itself.

Of course, not all employees will want to participate in new ways of doing things. This is a natural human reaction, and it is important to realize that the employees you have now are a product of what you have created, or allowed to be created. With no direction, no purpose, no instruction, and no principles, the culture may grow based on an employee's idea of how things should be, instead of the company's focus on how things should be. As you make changes, expect a bit of skepticism, but support the people who want to rise to the top. A meeting that is brief and to the point shows that the customer and professionalism is more important than gossip. There still needs to be time to have a discussion on various points in the meeting. However, there needs to be structure around it; otherwise, the line between work and play may negatively affect the meeting's agenda. As a business, you need to decide on the core principles and hold everyone accountable.

Principles are vitally important, as they provide the stability to support change. They let you provide direction without just making up rules or standard operating procedures to follow.

> Principles provide the framework for success and adaptation.

If we look back across the many years that survey research has been collected, while the overall numbers within each category change and the successful companies get recognized, the one thing that remains a constant is the importance of the relationship between an employee and their colleagues and supervisor. This relationship has the greatest impact on their overall happiness regarding their job. Your interaction with employees—showing empathy and caring for them, while at the same time holding them accountable—will be appreciated.

There are many instances in our careers when we personally have had to hold people accountable, and when our employees look back on it, they have told us they appreciated the accountability. Life is not easy. It is often a roller-coaster ride, so people appreciate a consistent and purposeful relationship at work. It is something they can depend on being consistent and fair.

Another large motivating factor that is consistently talked about is pay. Of course, everyone wants to make more money. However, surveys show that money-based motivation lasts only 3 weeks. At that point, employees feels that they are entitled to the salary, and more importantly, they have begun to build their lifestyle and spending habits around this new salary. What employees need more than anything else to stay motivated is an organization that separates recognition and celebration from people's pay. All too often, an organization does not have a sophisticated recognition system and compensation gets tied to the wrong elements of accomplishment and engagement.

For a good example of separating pay and recognition, we can take a look at athletes. They don't get paid more just because they were MVP of a season or scored the most points. These types of awards and recognition are frequently measured, so that many people over the course of a year could be MVP of a different game, or event, or the entire year.

As a child, Collin played ice hockey and remembers being frustrated by never being able to achieve MVP because another teammate was just that much better than he was. It wasn't until the other teammate was injured and pulled from a game that Collin was able to attain the status of MVP, when he finally had the opportunity to score more.

It was an unfortunate situation that allowed him to attain MVP for the tournament, but the idea is that enough measurements of MVP throughout the year will allow for circumstances to change, and many people can then be recognized for their hard work.

Some ideas we have encountered that help to recognize people in the business world are as follows:

1. Highest number of skills attained in the year
2. Most improvement suggestions made
3. Most replicable idea
4. Most valuable idea (chosen by the president)
5. Most innovative idea (chosen by the president)
6. Fastest setup
7. Best idea for quality improvement
8. Best idea for productivity
9. Most improved operator

10. Department that saved the most money (based on a formula tied to hours saved)
11. Team with the most ideas generated

These are all ideas that allow a business to celebrate people's ideas and action in a longer time frame, which is good, but an even more important system of recognition should happen on a daily basis.

Once again coming back to communication, this daily support happens through the sharing of information and handing off of responsibility to the next shift. As an example, Toyota does not stack shifts back to back. Instead, they have a 15-minute break between shifts so that production isn't simply passed on, and communication can happen before work begins again. This is key to showing that production goals are the responsibility of the team, and abnormalities or defects should not be passed on to others. It also highlights the integrity and respect for one's work and the work of colleagues.

RAISE YOUR VOICE TO RAISE MORALE

PEC has trained tens of thousands of people in true Kaizen through Muda-tori, and this process always begins with morale training. Because Kaizen starts by changing your own mindset and actions first, it is important that you begin training by focusing inward.

By focusing inward, you begin to train yourself to see what you can do or accomplish, and not what is happening around you. Your perspective will change, and you will begin to see things in a different light.

This was an important distinction that came about in Japan because of the necessity to go against the cultural norm of "harmony for all." Typically, in Japan, people are taught to embrace keeping harmony for everyone, and not just think selfishly about their own needs. For instance, you will find people stay within their own space on a train, don't speak too loudly, and are generally very cordial.

This needed to change in order for companies to achieve greater results. People needed to learn how to be disruptive to the normal day-to-day work, and that went against what they had been taught growing up. This meant that Japanese companies began employing people to train their workers to focus inward and break out of their comfort zones in order to gain confidence in this new way of working.

THE ELEPHANT IN CHAINS

A great illustration of the power of assumptions is a story written by Jorge Bucay, shared with Miura during his undergraduate studies. As a child, Jorge loved going to the circus. One day, he noticed one of the huge elephants was being held by a large chain around his foot, which was only fastened to a small wooden stake. This appeared to be a weak system, as the elephant was easily strong enough to simply tug at the stake and walk away whenever he wanted. So why didn't this strong elephant just break away?

Jorge decided he would ask, but every adult he asked simply replied that the elephant was trained. So with that explanation, he closed his eyes and imagined the elephant trying to get away, and it was not until he began running through scenarios that he had his breakthrough. He thought of the elephant as a baby, and the chain and stake would have been enough to hold a baby elephant in place, so the baby probably fought and fought to get away until it was too exhausted and fell asleep, before waking to try again the next day. This must have happened for countless days before, one day, the elephant simply resigned itself to its futility and gave up trying to get away. Because the elephant had come to the conclusion that it could not escape, even as an adult he believes he cannot escape, so a simple stake is enough to keep it in place.*

Because of the elephant's state of mind, even though the environment and situation has changed, he does not realize it. If we take this analogy to the workplace, how many times do we see situations change, but the complacency with the current state means nothing truly changes? Do your employees have the passion to break the chains, as the elephant used to have? You need to rekindle people's natural desires to break through convention and allow them to believe and see that Kaizen and identifying Muda (waste) are not only possible, but will lead to greater work–life happiness.

YARUZO CALL

As Miura has come to know, through his implementation of training, you can break through the walls and barriers we have all created around ourselves: the state of mind we come to call normal.†

* Jorge Bucay, *Dejame Que Te Cunte*, Del Nuevo Extremo, 1999.

† Hitoshi Yamada, *Hitome de Wakaru, Sugu ni Ikaseru, Kiso kara Wakaru Kaizen Leader Yosei Koza*, Nikkan Kogyo Shinbunsha, 2008.

One way to achieve this is a hearty shout in the morning. Try shouting as loud as you possibly can, from deep down in your gut. Shout at the top of your lungs by putting forth all your strength, shouting louder than you have ever shouted before.

Once you reach this point of breaking, you will leap beyond your initial assumption of how loud your voice can become. Generally speaking, people do not believe in their true potential and won't overcome challenges, assuming that it is beyond their power. To help them break their assumptions and get them out of their old molds in the shortest time possible, it is extremely effective to use the physical activity of shouting so that they can experience what it is like to overcome limitations. Remember that the first Kaizen that you always do is to change yourself and reset your negative thinking and behaviors.

This is where the concept of the Yaruzo Call comes from. The Yaruzo Call, or "Do It Call," helps you break down barriers and realize that you have more skills, more strength, and more experience now, and you can use that to reach greater, more enjoyable heights. Through its practice, you begin to break down the walls in your mind, and you will find yourself shouting, "I can break these chains so easily!"

This helps you to believe in yourself again, to believe you can achieve more at work, and it helps you to declare this to not only yourself but to your team members and coworkers as well (Figure 6.9). Everyone Miura teaches uses this Kaizen practice every morning and before Kaizen training events. Here is how the Yaruzo Call works:

1. Stand with your feet planted firmly on the ground.
2. Raise your fist with each line and shout, as loudly as you can:

"I will do it today!" "Do it, do it, do it!" "Do it first, then think!"

As you try this Yaruzo Call with your loud and energetic voice, while raising your fist high up in the air, you will realize that your actions will become faster and more responsive to your ideas. Based on Miura's personal experience as a Kaizen coach, morale training has been successful in boosting trainees' engagement in Muda-tori and generally speeds up their actions by over 5 times.

Morale training alone can change people's actions very dramatically. However, there are people who are convinced that their actions will follow only once their thinking has changed. In this case, we argue that there is

FIGURE 6.9
Yaruzo Call.

no clear indicator of when their thinking will change and what has to take place in order to change their thinking. Kaizen is about taking action, not who has the best theories or ideas. You execute your own ideas by taking action. You achieve results because of these actions, not because you have thought about taking action.

Miura often asks trainees, before going through morale training with them, how they determine if someone has a strong "can-do" spirit and willingness to try new ways. About 70% of trainees think it is represented in the loudness of one's voice. Around 20% of trainees believe that it is represented in the speed of one's actions. The remaining 10% of trainees responded that it is represented by a glint in one's eye, an intangible feeling. Usually, one's motivation can be explained by a mindset or feeling, but in the case of true Kaizen, it is assessed by the loudness of one's voice and briskness of one's actions. Only through actions can you demonstrate the mental changes of Kaizen.

The state of one's mind will surface through their words and actions. Therefore, in reverse, it is true to say that correcting actions and the loudness of one's voice will put one's mindset in an excellent condition (Figure 6.10).

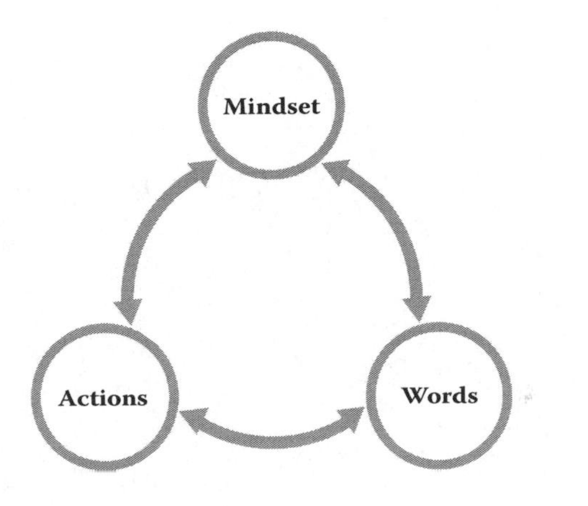

FIGURE 6.10
Mindset, actions, words.

After training more than 10,000 people in morale at PEC, Miura discovered that it takes only 1 minute to change the loudness of one's voice and actions, whereas one's mindset cannot be changed right away. During morale training, the focus is on reenergizing trainees by encouraging them to put forth all their strength in shouting and moving their bodies at a brisk pace. Muda-tori is all about yielding results as quickly as possible. Good results will lead to a greater confidence in ourselves and inspire us to keep executing Kaizen in order to achieve even better results.

CONCLUSION

Morale can be difficult to assess, but by examining the physical status of your facilities, the quality and quantity of your employee interactions with each other and with your customers, and the displays on the walls, you can begin to understand where your organization stands. By improving these factors, you can create a more alluring and enjoyable work environment, and you can communicate your organization's values to the greater community.

7

Identifying and Removing Waste

Looking back, in Chapter 6, we discussed assessing your facility and explored the ideas and contributing factors toward morale, and in earlier chapters, we covered the various types of management systems. In this chapter, we examine waste in all its forms: what waste looks like and the things you can do to eliminate it.

Whether it is learning to observe and identify waste on the factory floor, or empowering your employees to remove it from their processes, the elimination of waste should be central to your improvement efforts. However, in order to eliminate waste successfully, people have to be sensitive to the wastes in order to see them for what they truly are. It is important that an effort be made to hone employees' skills in seeing the wastes that are present and supporting them so that they embrace the idea that the status quo is not good enough. They must learn, experience, and be supported by training in order for waste elimination to be successful and people to want to use their efforts to remove wastes. This is also important because employees can then focus any downtime on waste removal, instead of it being idle time.

According to an article on the amount of time wasted at work, posted by SFGate,* 69% of workers responded that they slack off at work; 24% say they waste up to 60 minutes a day doing activities unrelated to work. That adds up to 5 hours each week that is not productive, by at least 24% of the working population. A separate survey, completed in 2012 by thegrindstone.com,† points out that workers complete only about 29 hours of work per week.

* 2014, SFGate.com http://www.sfgate.com/jobs/salary/article/2013-Wasting-Time-at-Work-Survey-4374026.php

† http://www.thegrindstone.com/2012/05/31/career-management/you-are-actually-only-productive-for-29-hours-per-work- week-706/

The magnitude of these statistics seems astronomical, yet over time, the statistics have not changed much and these findings are not unique to a particular industry; these statistics are found throughout every segment of business. This is where the true opportunity lies to connect employees to organizational strategy and purpose so that they are sensitive to the purpose of waste reduction as well as the fact that their actions are related to organizational strategy and the organization's purpose.

> 29% of people say they waste up to 60 minutes a day on non-work-related activities.

DEFINING WASTE

Wasted time is just one segment of possible wastes. We must broaden the question: what constitutes waste? When business people respond to the question, their response typically gravitates toward idleness, personal time on the phone, computer games, unproductive meetings, and disruptions. Very few (almost none) challenge the lack of value-added content that happens in various activities.

When non-value-added activities are added to the definition of waste, the amount of time spent on wasteful activities can increase to 60%–70%. In other words, it is not that employees are not busy, but instead that they are occupied with tasks that do not add value in the eyes of the consumer.

Waste should be defined as non-value-added activities or outcomes. Typically, activities that contribute to waste are defined based off of Taiichi Ohno's model of the 7 Deadly Wastes:

- Overproduction, or producing items before they are needed
- Transportation, in and out of storage
- Waiting, for material or information
- Motion, unnecessary or excessive
- Processing, either in repeating operations or information
- Inventory
- Poor quality, lots of defects

And often, the eighth waste that has been added onto the list, credited to Norman Bodek:

- Not tapping available human capital/potential

When going through training, many managers who begin to use this definition of non-value-added waste invariably determine that around 70% of all time spent is non-value-added. However, the real amount of waste is actually higher, because there is an opportunity cost associated with waste that most people don't account for. For instance, the time taken up by non-value-added steps could have been spent processing items that are actually needed. Capital that is used to pay for storage facilities to store inventory instead could be used to fund efforts to make the workplace more productive. Within industry, well over 50% of all cost related to meeting customer demand is administrative in function, meaning it is nonproduction related and, therefore, a non-value-added activity from the perspective of the customer.

Given that a great portion of the economy is in the service industry, it is important we challenge waste in office functions. Although one response may be that a service organization does not have the problems of waste that occur in factories, there are a number of tools that can be used to determine the potential for waste reduction in the office. For an office environment, such techniques can be used to reveal the hidden wastes for people. These are as follows:

- Visualize the process
- Analysis done where the value is created
- Use words and graphical illustrations
- Make sure everyone is personally involved
- Use facts and use information

Remember that the blame is never with the people, but rather the processes and environment in which they work. Management consultant W. Edwards Deming* claimed that 95% of the variability in performance should be addressed to the system or process design, and only 5% of the performance is due to how well the employees apply their skills. Although counter-intuitive, if true, then first focus on the system.

* http://en.wikipedia.org/wiki/W._Edwards_Deming

Four basic techniques can be used to quickly assess the state of operations.

Disassemble First

In one factory Miura coached at, he asked the president to disassemble one of the finished hydraulic motors and reassemble it. This would give an accurate timing on how long it could be assembled, instead of assuming assembly took the full 4-week lead time they were currently at. In reality, once he had the motor disassembled, it only took 1 hour to reassemble.

Focus on looking at the end process and working backward to get your ideal lead time, which is the sum of the operator cycle times on a particular part.

This is the point where Taiichi Ohno would have torn down the jig racks, taken out inventory, gotten rid of computerized machines, and gotten down to the core idea of flowing work with no work-in-process (WIP). This kind of thinking, that it is okay to have a 4-week lead time, allows non-value-added portions of tearing down and building up work to be accepted when, in fact, this is one of the issues we have the most control over.

We encourage you to think about this from the shipping dock back upstream, or if you don't ship product, then from the latest point in your processes before the final interaction with your customer. Think about how long it would take to satisfy real demand. Reduce the distance created by the use of computerized machines, racking systems, and transportation systems. If all of it was gone, you could connect and respond to your customer's needs right this minute.

Green, Yellow, Red

Inventory's only purpose is to serve as a time buffer for unbalanced activities. Therefore, having inventory is a technique to run areas that are not as synchronized as they could be. We use the Post-It Notes technique to highlight excess inventory. With three colors (green, yellow, red/pink), start labeling all the inventory. Use green for items to be used in a few hours, yellow for a day or two, and red for longer times where inventory is not moving (Figure 7.1).

Once everything has been labeled, everything that is not labeled in green should be removed from the work area.

FIGURE 7.1
Green, yellow, red stock level lines.

The reality is that, in most places, the majority of your labels will be red, and you will discover hundreds of items that can be eliminated or at least stored when not in use.

Spaghetti Diagram

Every step taken costs money, and every step not taken is money saved. To use the Spaghetti Diagram (also known as a Motion Diagram), have someone follow a worker around for a few hours and trace the person's pattern of movement onto a map of the facility. When the time is finished a typical drawing will resemble a plate of spaghetti, with overlapping lines (Figure 7.2).

Assume each step taken costs an average of two cents. Therefore, by simply moving pieces and parts closer to each person, you can save hundreds of dollars every week.

Activity Sampling

As a manager, you need to be aware of every time someone steps onto the actual floor, be it an office or production area. Start tracking every time an individual enters the work area. What kind of activity are they involved in? By categorizing value-added and non-value-added activities, you will begin to find where time is being wasted.

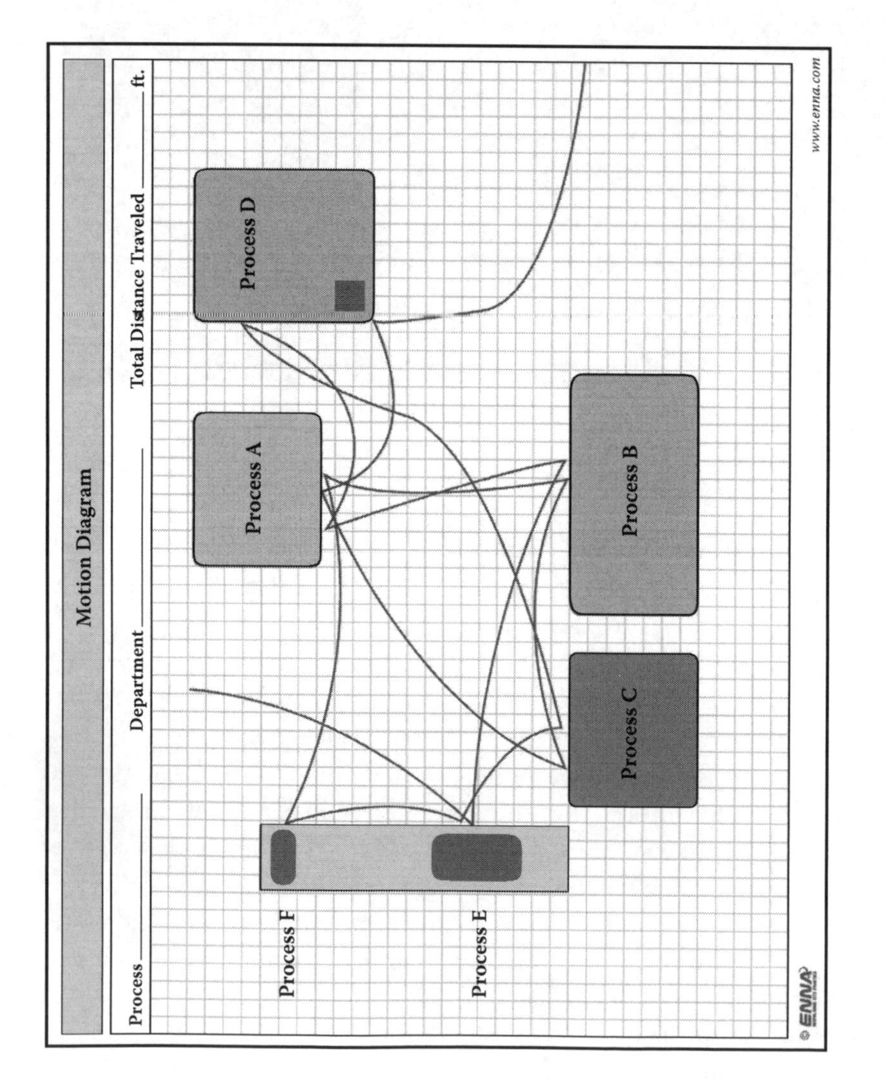

FIGURE 7.2
Motion diagram.

This approach should always be used to mentor managers to learn to see work and waste; to begin to build their own sensitivity of the conditions that cause additional costs.

REMOVING WASTE ONCE IDENTIFIED

Once you have opened your eyes to the waste in your facility, it is time to work on eliminating it. One key factor, which differentiates successful teams from others, is personal discipline. The clearest evidence of personal discipline is reflected in the workplaces of the individuals in the work area.

On one of Collin's early visits to Japan, Sinko Air Conditioning's operations manager's first statement to the group was, "In this plant, we walk at so many paces per minute." As we mentioned earlier, they use a morning exercise to set the pace for their facility, and then the workers maintain that throughout the day. The implication began to sink in with Collin after the first few minutes of witnessing this. Intentionality: "where am I going and for what purpose?" should be reflected in the level of activity.

Intentionality: Where am I going and for what purpose?

IMPLEMENTING THE 5S's

One practice indicative of employee engagement is the application of the S's (be that 2S, 4S, 5S, or 6S) and the extent to which waste is removed from operations. This Japanese practice is a popular choice for early implementation of Kaizen because it is so visual in nature. When you break it down, the importance of 5S is that it helps you to organize your space and remove waste. The ability to maintain the 5S's is the best indicator of the discipline required from plant management (Figure 7.3).

Traditionally, you can observe the first 4 S's and then make sure you keep up with them (the fifth S). Modern interpretations have shortened this to 2S (Cleaning and Self-discipline), but the basic principles still hold.

FIGURE 7.3
Breakdown of the 5S's.

MAINTAINING 5S

It is not enough to just have the 5S's practiced in the maintenance of equipment and the condition of the floor, which need to be manually cleaned. 5S must be implemented throughout the organization in order to have a cultural effect.

By using the process of cleaning, and maintaining that cleanliness, you are designing processes for employees to succeed. Make sure everything is in working, operational order for employees to be able to do their job well. This will ensure work is safer and easier and generates a higher-quality product.

When we conduct an assessment on factory equipment, we first need to examine how well cleaning is currently being implemented. Cleaning is the foundation for everything, including Kaizen. As an example of the importance of cleaning, consider a well-maintained restaurant kitchen. They are always kept impeccably tidy, and the 2S's (Sort and Set in Order) are well sustained. Only the ingredients to be used are put on a chopping board, exactly when necessary, and are always kept fresh for preparing and cooking. The tools the chefs use are well maintained, especially the knives. Customers do not continue to patronize a restaurant whose kitchen is not clean and tidy. Even when the restaurant business is thriving, success may be temporary for a business if the kitchen is not kept clean.

So how can we apply this idea of cleanliness to a production shop floor? Begin by looking at sources of uncleanliness. Drops of dirty oil around machines can lead to sloppy work as well as safety concerns, such as workers slipping and falling. Therefore, it is extremely important to clean machines and tools in a comprehensive manner prior to implementing Kaizen.

In Japan, cleaning is part of the day-to-day culture, and all students in school, from elementary school through high school, are taught to clean their own classrooms every day. Because of this, employees in Japan are also expected to clean each of their workplaces on a daily basis.

The idea is that if everyone continues to maintain his or her workplace through cleaning, a higher quality of work will surely come out of it. Cleaning is the basis for increasing awareness of oneself. As you sweep and polish the floor by wiping it with a scrubbing cloth, you will begin to see more and more dirtiness that is revealed underneath items that often go overlooked. As more cleaning takes place, you will notice when there is a piece of trash on the floor, and soon your automatic reaction will be to pick it up and find out why it happened in the first place.

With this change in thinking, you will start picking up trash from the shop floor, no matter how small it is, in order to maintain the level of cleanliness. Then, you can begin applying your creativity to prevent even a small piece of trash from falling to the floor. Don't limit your cleaning only to machines though. Work toward improving the level of cleanliness in the entire workplace in order to strengthen the capacity of each work area.

Once you've removed some of these tangible sources of waste and wasteful activities, you can begin examining the flow of materials through your facility.

WASTES OF STAGNATION AND MOTION

PEC divides waste into two categories: Stagnation and Motion/Transportation. Stagnation focuses on items, when they are proceeding through the facility and when they are at rest. Motion/Transportation analyzes how people move while they perform work.

Miura's sensei, Hitoshi Yamada, was the first person to have articulated the concept of Stagnation waste, back in 1975 when he was learning about the Toyota Production System directly under Taiichi Ohno. Mr. Yamada learned that the ratio between the process time and stagnation time within Toyota's most efficient facility was 1:300.

That means, for every 1 minute a product was being worked on (in process), it sat waiting for 300 minutes. Mr. Yamada was surprised to find that even Toyota, the world leader in providing products just-in-time, had so much stagnant time.

To clarify, when we talk processing time, we are talking about the actual time spent creating value for the customer. For example, in a stamping process, the processing time is only 1 second: when the giant dies pound together, turning the raw materials into a stamped piece of metal.

The time in which each die travels back to its position and the raw materials are inserted is not calculated in the processing time because the item in question, in this case the raw piece of metal, is left stagnating—simply a work left between value-added processes.

The time spent for preparation of raw materials or unnecessary movement of machines does not add extra value for the customers, so it is considered Stagnation waste.

With this understanding of processing to stagnation in Toyota, Mr. Yamada went out and began to gather data at other facilities in Japan to see how they compared. His findings showed that companies that were making money had a ratio of 1:5,000, whereas companies that were losing money had a ratio of 1:10,000 processing to stagnation time (Figure 7.4).

Production lead time = **Processing time** + **Stagnation time**

Company comparison	Processing : stagnation Ratio
Toyota	1 : 300
General companies	1 : 5,000
General companies in deficit	1 : 10,000

FIGURE 7.4
Stagnation waste to processing time ratio.

At the time of his research, Mr. Yamada was focused on increasing productivity by reducing the processing time and implementing cutting-edge automation systems. After all, that was how American companies had been so successful for many years, so why not implement the same advanced automation systems to mass produce items?

Soon though, Mr. Yamada realized that he needed to focus on eliminating the WIP, as they were key to the Stagnation waste existing. This was how he could make various companies profitable once again.

Mr. Yamada blamed various inventories found on the production floor to be the main cause of prolonged wait times and labeled these inventories of stored goods (whether they were WIP, raw materials, or finished goods) Stagnation waste.

Every time Mr. Yamada coaches in a factory now, he begins with the identification and removal of such Stagnation waste. Eliminating Stagnation allows companies to reduce their stored inventories and increase their cash flow by freeing up all the resources currently being tied up in WIP items.

HOW CAN WE IDENTIFY STAGNATION WASTE?

The value of inventory assets as described on your corporate financial statements include the values of finished goods inventory, half-finished inventory, WIP inventory, and raw materials. However, you don't see bags of cash sitting on the floor in your facility. Instead you see finished goods and WIP items on pallets and carts. These are all part of the waste of Stagnation.

Put simply, the most effective way to reduce Stagnation waste is to completely eliminate storage shelves and reduce the number of pallets and carts in the business. This limitation forces you to think differently about how you produce your products.

A few years ago, Miura worked with a furniture company, Maruni Wood Industry, which had embarked on their journey by first removing the conveyor belts used for their packaging line. By bringing their processes closer together and creating a more flexible space within the factory (a process called "Majime"), they were able to reshape how they handled time management.

They created a flow between different processes, and their production lines were improved through Kaizen in order to accommodate a

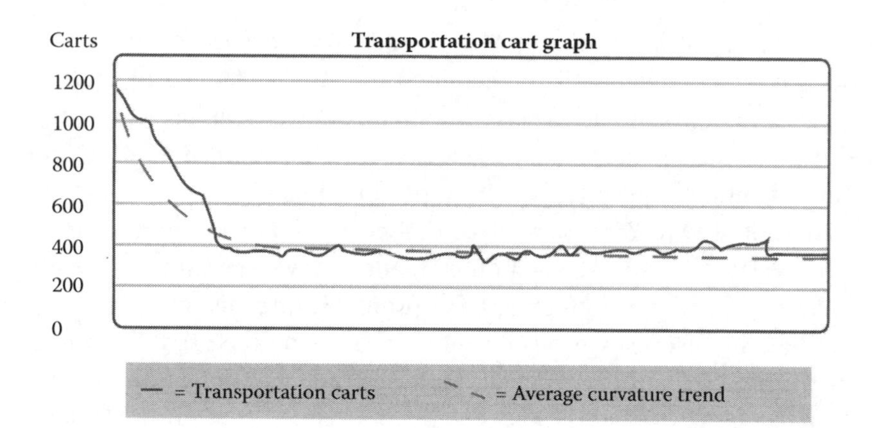

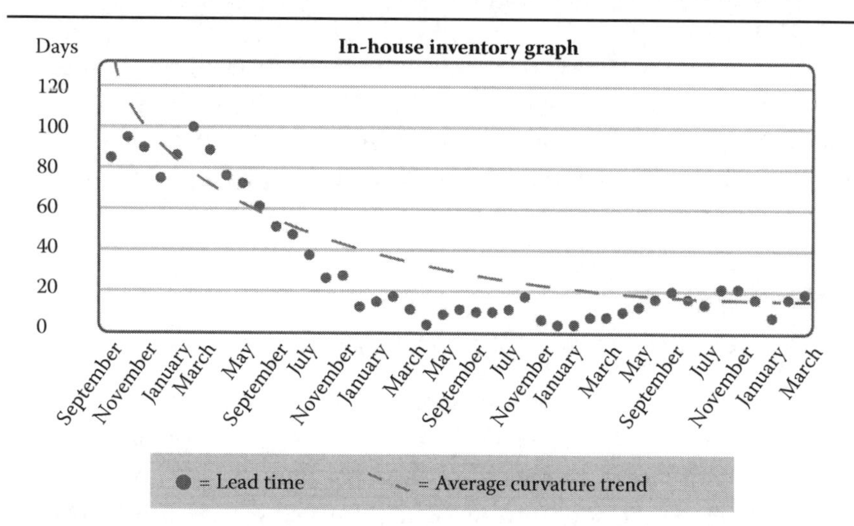

FIGURE 7.5
Transportation carts and finished goods inventories at Maruni Wood Industry.

> MAJIME:
> The process of bringing items closer together.

small-lot production system. As a result, the number of transportation carts between steps decreased over a 2-year span from 1200 carts to only 400. Their entire production system has since shifted to a complete make-to-customer-order production system, reducing their finished goods lead time from 3.5 months to 10 days* (Figure 7.5).

* Hitoshi Yamada, *Forging a Kaizen Culture*, Enna, 2011.

THE STORE AND THE FRIDGE

Many people say they have implemented a Lean system, such as 5S or 2S, in their business and have achieved great success, but if you take a tour through their shop floor, it is likely you will find hundreds of items that are probably not going to be used that day for production. If you ask one of the employees if a tool or piece of material is necessary, they will probably say yes because it could be useful at some point, not because they need to use it that particular day.

To help workers decide if something is necessary, you need to provide them with a time frame so they can appropriately determine the necessity of an item. Instead of asking if something is necessary, try asking, "Exactly when do you need this item today?" Only items that will be used that day should be left in the process work area. If a tool, material, or piece of equipment is not being used that day, it should be put back.

> To determine necessity of items, ask:
>
> "Exactly when do you need this item today?"

We call the area where items needed today are kept the "Fridge." The area upstream of your process where you return unnecessary items is called the "Store." PEC uses these concepts of the Store and Fridge to determine the correct placement of necessary items in order to eliminate Stagnation (Figure 7.6).

In the Toyota Production System, they call the area where finished items are placed temporarily the "Store." Mr. Ohno taught that the location of a Store should be where items are being finished. Workers from the downstream process pull the items, in necessary quantity, from the Store location, then downstream workers place those items at the beginning of their production line, or the Fridge.

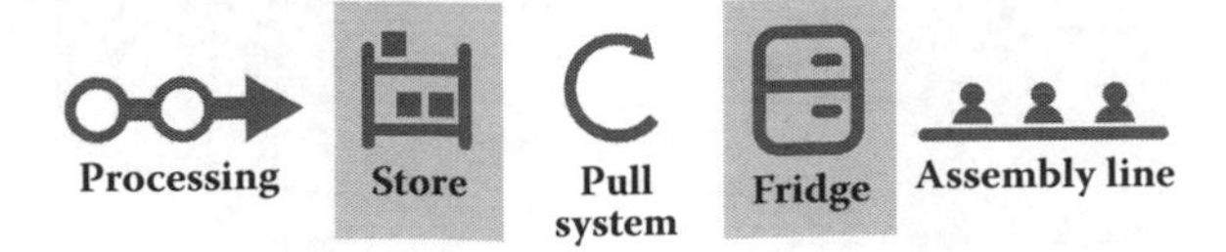

FIGURE 7.6
Illustration of Store and Fridge locations.

It may be helpful to think of this in terms of a household setting example. You go to a supermarket/store to purchase the necessary ingredients you need for the next few days or weeks and bring them home to be stored in your refrigerator.

Back at your work processes, if there are too many items in your Fridge, you need to think about removing every item from the Fridge that is not needed for that day. If there are too many items in your Store, you are overproducing and need to examine your upstream processes. This way of placing items, by using the concepts of Store and Fridge, helps easily identify Stagnation waste, no matter what people's level of Kaizen understanding is (Figure 7.7).

It also helps if you think about each process within your facility as a separate customer. Items that are placed in a Fridge are what people from the downstream process have "purchased."

There is no way to eliminate the waste of Stagnation if your employees continue so stick to the point of, "We have produced exactly to what the production department ordered us to do. Therefore, we have not overproduced" or "When the items will be used in downstream processes has nothing to do with our internal process flow." This is a mindset that must change in order for Stagnation to be eliminated.

Every employee must understand what the most efficient level of production at your facility is. By utilizing the space on your shop floor with a Fridge and Store, you can quickly illustrate to everyone what the proper amount to be produced is.

	Store	**Fridge**
Placement location	Where items are produced.	Where items are used.
How items are placed	You know who the items pull. Where and when the item was produced.	You know when the items are to be used.
Management	Workers who produced the items. Workers who purchase the item.	Workers who use the item.

FIGURE 7.7
Placement of items in the Store and Fridge.

GIVE YOUR MATERIALS THE 3F

Another tool for reducing Stagnation is the 3F Management concept. This concept, used for organizing materials, stands for Fixed Location, Fixed Item Identification, and Fixed Quantity.

Fixed Location: Each item needs to be placed in its designated area. Lines are drawn on the factory floor to clearly separate the areas of Store and Fridge. These locations do not change—they are fixed.

Fixed Item Identification: Items are always placed in the same location. As long as each different type of raw material is always placed in the same fixed location in the raw material Store, when workers come to "shop" on a daily basis, there is no need to label anything—they simply look in the same place as the day before. This reduces time wasted on searching.

Fixed Quantity: Only a certain amount of items are allowed to be placed in a space, often by making it physically impossible to fit more items there. This reduces the wastes of overproduction and overpurchasing.

The elimination of Stagnation waste always starts with visualizing how items or materials are placed. This allows managers and their employees to quickly detect if a situation is normal or abnormal and gives time to prevent errors from occurring.

USING KANBAN TO ELIMINATE STAGNATION WASTE

A confectionary producer in Japan, Suzette, which sells to various department stores across Japan, has built their business model on providing high-quality sweets to customers at a reasonable price. They are able to offer such deals because they have achieved significant cost reduction in their production facilities through implementing true Kaizen.

To maintain their competitive advantage, they have also started implementing the same Kaizen concepts into how they treat the retail stores their products are in. Their first step was to focus on managing how each item is placed and packaged in each location.

Each retail store had a different minimum stock level for the packaging material used to wrap the confectionery. This was an important point to standardize and systematize, due to the cost of labor and packaging products. This variation in the on-hand stock of packaging materials caused some retail stores to keep as much as 2 weeks worth of packaging materials for potential future orders from customers. They needed to eliminate this waste because the storage of the packaging materials took up a lot of the store's space and locked money into those materials.

Their solution was to work with suppliers to change the lead time required for packaging materials to be delivered to the specific store. Through working together, they managed to bring down the time so that when retail stores wanted to order new packaging material, it would now be delivered in 1 day. To build on this success, a Kanban ordering system was implemented.

In this situation, each Kanban represented an order card to their suppliers. To make the ordering process easier, each type of packaging material was given a tag that contained a unique ID number. For instance, when the wrapping paper labeled Kanban Tag 1 was used by a retail employee, the tag was given to the retail manager. At the end of the day, when the business closed, the retail manager would order that particular wrapping paper from the respective supplier, and the item was delivered to the retail store in the early morning of the second day after the order was placed. (If ordered on a Monday evening, the packaging arrived by Wednesday morning before the store opened for the day.)

This Kanban system allowed them to order necessary items in a much shorter lead time so that they only had to carry one day's worth of inventory in their retail stores (Figure 7.8).

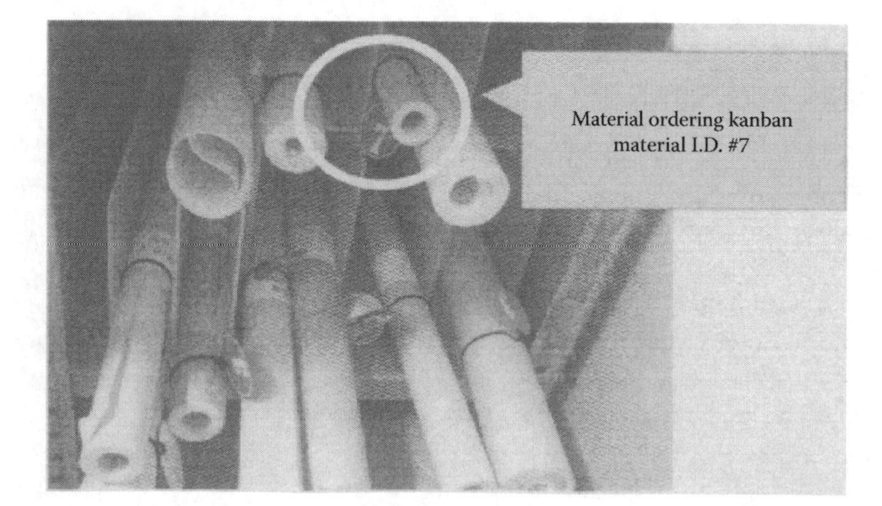

FIGURE 7.8
Kanban ordering system.

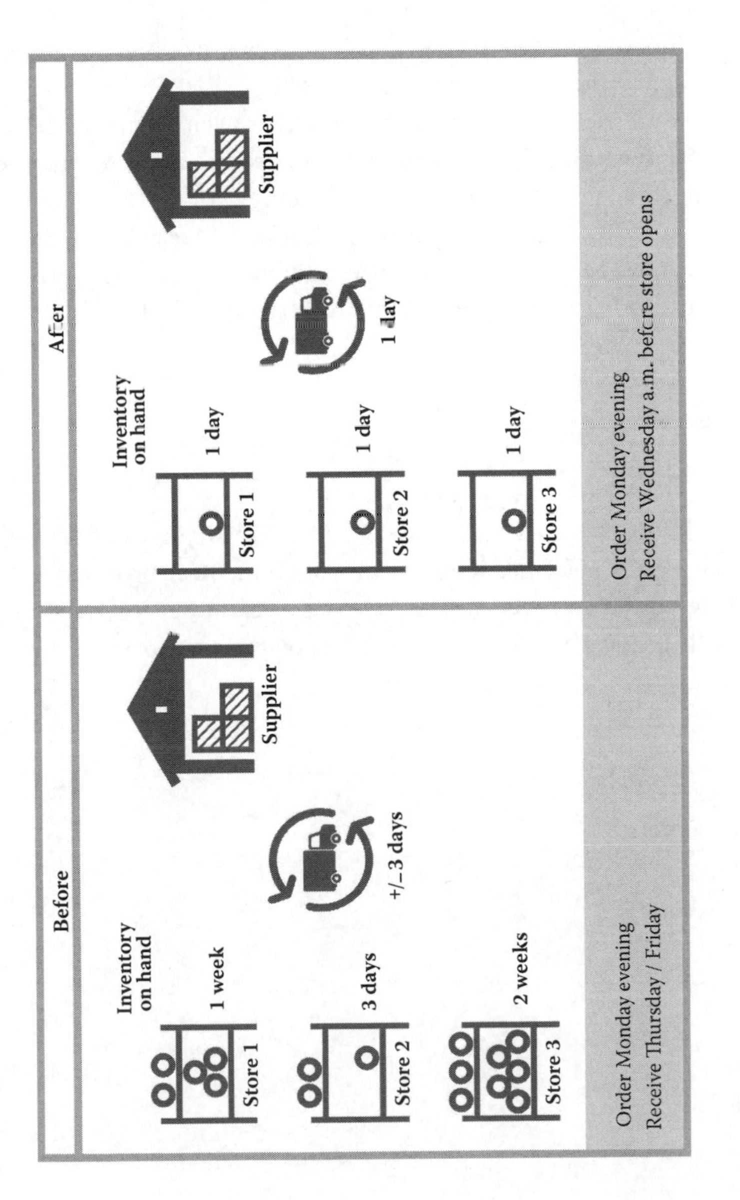

FIGURE 7.9
Before and after Kaizen—Kanban ordering system.

Currently, the inventories in their retail locations do not exceed more than 3 days, and the material ordering process in the stores has become extremely easy to manage because of the implementation of their Kanban system. Ms. Aya Sato, the retail manager from their Yokohama store, proudly told us that, "Periodic stocktaking tasks now only take 15 minutes, compared to the 1 hour it took in the past."

This Kanban ordering system is currently being implemented at over 50 other retail locations. Executive director of production, Mr. Masanobu Tomita, reports that the overall material inventory of all the retail stores has been reduced by 21.5% so far. It is Mr. Tomita's forecast that the total inventory will be reduced more than 50% after every retail location across Japan has successfully implemented this Kanban system (Figure 7.9).

CONCLUSION

By implementing concepts such as a Store and Fridge, you can begin eliminating the wastes you have identified throughout your production line. By challenging the assumed needs your organization currently operates at, you will discover waste in all forms, allowing you to use Kaizen to eliminate it.

8

Using Production Boards for Time Management

As you begin eliminating waste, you will naturally start to ask yourself some questions:

How do we know what the right amount of production is?
What is an appropriate lead time to set between processes so that everything flows evenly and no waste accumulates?

To begin to answer these questions, there are a number of steps you can take to make everything within your processes visual. This includes how the day-to-day schedule is measured, how you schedule your day as a manager to ensure everything is running smoothly and on time, and how maintenance is performed to ensure you are at, or nearly at, 100% machine uptime.

DAILY CHALLENGES OF A MANAGER

Each day, a manager faces an array of challenges. Typically, a manager's immediate concerns revolve around these factors:

- Preparing for the day's/shift's activities
- Analyzing and updating schedules
- Changes that occur throughout the shift
- Special circumstances (defects, rework, requests...)

All of these are important to understand and focus on each day. However, once you adopt a Kaizen mindset, you will be able to change your focus, as these will be easily taken care of each day. By making production information readily available to all workers, through the Store and Fridge concept, you can reduce the need for goal setting on a daily basis. Similarly, through the use of hourly production management, scheduling is already handled so you won't need to spend much time on that aspect of management and can focus your efforts elsewhere.

So where should you be focusing on a daily basis? With Kaizen adopted in your organization, you should begin focusing on:

- Producing what is needed, when it is needed
- Tracking hourly results in each work area, inventory, and turns of inventory
- Tracking Kaizen ideas and their implementation

You can see the shift in focus from the first list to the second. Instead of simply going through the daily grind, your energy as a manager can be spent on assessing progress and striving for improvement. This is the coaching mentality. A head coach of a sports team doesn't spend their time focusing on telling everyone all the details of the daily goal (winning the game) because everyone knows the goal and is working toward achieving it. Instead, the coach can spend time observing team members, checking that they are learning from practice, and coming up with fresh ideas for improvement.

PRESTART ACTIVITIES

You want to start each day with precision and commitment. We've already discussed the power of a personal greeting from the supervisor or manager and the effectiveness of a quick stand up operational meeting to bring everyone up to date on the latest production news. These techniques foster a constructive relationship between team members and management and nurture a feeling of being part of something larger than oneself.

The quality of interpersonal interaction makes the difference between being an individual player and being a team member. From a daily operations point of view, the first thing to do is focus on the aspects of the operation that are within your operational control.

Prior to start, a swift walk around the factory, can give a lot of insight. It is often at this point that you can identify any counterproductive practices, and correct them, even before production begins for the day.

MANAGEMENT'S DAILY STANDARD WORK

Managers must prepare daily for production, and this includes the startup and communication processes. The purpose is to get an understanding of the needs of the business and past actions. What we have found that a good approach and sequence to the daily work is the following:

- Review the previous shift's production and review any problems or potential problems.
- Verify key areas, such as machinery status.
- Check the availability and quality of raw materials, parts, and consumables.
- Review absenteeism and other workforce requirements, including the skills necessary for the day's production.
- Allocate personnel and update supervisors and team leads of any changes.
- Welcome employees as they arrive at work.
- Ensure work is ready to go so that business begins right on time.
- Walk through the plan for the day with your team and discuss any particular needs or areas in need of attention.
- Attend brief team meetings conducted by supervisors addressing any daily production issues and updates from the company.

TO PREPARE, YOU MUST HAVE DAILY DATA

An essential part of being prepared to effectively manage an operation and allow others to manage as well is to have accurate, real-time, and specific data at all layers of the business. If a business focuses on this kind of data, then the meetings and communication throughout the company can be quick, agile, and actionable. Doing this allows each supervisor or manager more time and opportunities to share with parallel departments or upstream and downstream departments to synchronize operations and information.

A useful Lean tool to use to convey any issues or problems visually to the entire workplace is the Andon. An Andon is usually a light, placed on a machine or production line, used to signal normal or abnormal processes. Typically, the lights are color-coded as green (normal), yellow (warning that something is imperfect), or red (abnormal, the machine or line is stopped). You often hear of Toyota's Andon System, in which a person working on the production line has the opportunity, and obligation, to pull a cord to turn the system to yellow or red to request assistance and even stop the entire production line.

With tools such as this in place, you can quickly and efficiently gather the necessary information on your production processes in real time. The more current the information is, the more timely corrective actions can be taken.

EVERYONE CAN CONTRIBUTE TO TRACKING DATA

It should not be the sole responsibility of managers to gather and communicate information to the workplace. Every employee should bear the responsibility of making operations more transparent. An example of how everyone can contribute to data gathering and communication across the organization is to take a look at the following example of a department store in Japan.

A major department store in Japan posts one person at the entry point, and it is their job to not only greet shoppers but also keep track of the number of people entering the store, using a handheld counter. By the end of the first hour, they have a good idea of how busy the day will be, based on actual customer flow data. This information is then conveyed to the sales people on the floor, so they can judge the traffic in their department and request additional staff if necessary or reassign team members to other tasks such as restocking. By gathering information early, they are able to be flexible, allowing managers to make decisions based on the current evolving conditions.

By ensuring everyone has access to important current condition information, you can minimize second-guessing, allow your workers to prioritize and organize, and assess practices to stay on the same page—all because everyone has access to the same set of data.

VISUALIZING THE DATA

Adding to the idea that everyone can and should contribute to gathering data is that this data should be made visual and easy for anyone to understand. Information such as daily customer orders, production needs, and product mix should be easily visible to everyone in order for production to run smoothly.

This open knowledge will also allow you to begin scheduling your days in a different, more efficient, manner. But how do you begin to interconnect your various processes so that they all flow together to achieve smooth production? One method is to start with visual management boards.

A visual management board should be installed on the shipping dock, as this is the most critical point in your process because it is the last contact point before the customer uses your product. This board should clearly show the daily shipping schedule, which allows everyone to see which trucking company is picking up which particular items at each given time of day. As long as every scheduled outgoing shipment is checked at the end of the day, it means that finished products are correctly on their routes to their respective customer.

Let's examine an example of a visual management board for shipping at Kashiwa (Figure 8.1).

If, in this example, it was 2 p.m., the board clearly tells us that the shipment to be picked up by Logistics Company C is not yet finished but is due (looking at the "Packaging Prepared?" column), meaning this shipment will not be made in time to make it to the end customer on time.

Although the argument could be made that it is possible for the shipment to be packaged quickly prior to the truck arriving, if workers had realized that the load was not packaged by 1 p.m., they would have had time to package it in a reasonable time prior to the truck arriving.

Visualization of critical information, such as shipping schedules and production progress, will increase the number of people who can identify

Shipping management board - February 15 (Wednesday)					
	Logistic company	**Product name**	**Quantity**	**Packaging prepared?**	**Delivery completed?**
8:00					
9:00	Logistic company A	Product E	1000	Done	Done
10:00					
11:00	Logistic company B	Product F	1500	Done	Done
12:00					
1:00					
2:00	Logistic company C	Product G	1200		
3:00					
4:00					
5:00	Logistic company D	Product H	2000	Done	

FIGURE 8.1
Example of visual management board for shipping.

issues before they become critical and respond by taking necessary actions to solve them.

VISUALIZING PRODUCTION MANAGEMENT

The purpose of visualizing production is to show everyone whether work is running ahead of, on time, or behind the schedule. If we find out that we are behind schedule at 5 p.m., it would be impossible to recover because

Management is to help us determine what is normal and abnormal, quickly and accurately, at any given time.

Sensei Yamada defines true management as follows:

"True management" is to

- reveal what is normal and abnormal.
- reveal what is early and late.
- reveal what is good and bad.

Visualize a situation to everyone so that they understand the state of the situation in the most concise manner.

'Kaizen tamashi no sakebi' Yamada 2011

FIGURE 8.2
Visualizing production management.

everyone is eager to leave work and go home. The sooner you find out if work is falling behind schedule, the easier it is to catch up. The quickest way to eliminate delays in work is to respond immediately to remedy any delays, instead of putting it off to the end of the workday (Figure 8.2).

A production management board must indicate critical information such as what items need to be produced and how many items must be completed by a certain time throughout the day. At every interval, the actual production achieved needs to be written next to the goal for that particular time period.

If there was a delay, causes and explanations should be documented on the board. When production is running as scheduled, this means that it is being managed without any issues. However, if there is the slightest delay in production, this written documentation shows how it was immediately solved and addressed (Figure 8.3).

It is the role of managers to frequently visit each shop floor and understand what types of existing issues have been encountered or solved. It is also important to work with the people in each work area to ensure the same issues do not occur again, especially in certain production lines that are vulnerable to particular types of issues based on their history.

In cases where there are many issues with machinery or equipment, a maintenance department should be alerted so that they can perform more extensive care on particular machines. If defective raw materials are the cause of a problem, the purchasing department needs to be provided with the necessary information to educate their suppliers.

Production management board for the production line A—February 15				
Hour	**Type**	**Goal**	**Actual**	**Issues/ solutions**
8:00–9:00	Product A	100	100	
9:00–10:00	Product A	100	90	Production line stopped for 7 minutes due to machine breakdown
10:00–11:00		100	110	Work caught up by adding one support worker
11:00–12:00	Product F	90	90	
12:00–1:00		90	90	
1:00–2:00		90	90	
2:00–3:00	Product G	120	100	Raw material defects (20 units)
3:00–4:00		120	100	
4:00–5:00				Work caught up by adding one support worker

FIGURE 8.3
Production management visualization board.

Last but not least, managers must fully engage with all other supporting departments in order to solve issues that are beyond the control of the production department.

TRANSITIONING FROM DAILY TO HOURLY MANAGEMENT

Production floors that are managed on a day-to-day basis naturally manage the traffic between processes in 1-day increments. This means that production of an item that uses three processes should have a lead time of 3 days. However, to eliminate the waste of Stagnation effectively, you need to synchronize your various production processes. The best way to do that is to transition your production from a daily to an hourly management system (Figure 8.4).

Start by managing your production by half-day increments. For example, the first process would be completed by noon, and the next process would be able to be performed in the afternoon. This allows you to cut your lead time in half. The next step is to manage your production by 2-hour increments, which allows you to cut your lead time in half again. If you have a process that takes longer than 2 hours, simply schedule that process for multiple 2-hour blocks of time, but make sure you keep everything else moving at that rate.

As you begin to break down your scheduling from full days to hourly increments, it may seem impossible to meet the new goals. However, as people begin to get close to the goal they will begin to find ways to cut down the lead time, whether that is by reducing waste, better maintaining equipment so that there is less downtime, or rearranging processes to accommodate easier movement.

Better time management not only eliminates wasted production efforts, it is also essential for boosting productivity. Try implementing a time management system for your production processes in order to synchronize the various processes in your facility.

After you get better at eliminating Stagnation wastes, you will become able to more effectively manage your production by progressively smaller

Management	Process 1	Process 2	Process 3	Lead time
Daily	Day 1	Day 2	Day 3	3 Days
Half day	Day 1, a.m.	Day 1, p.m.	Day 2, a.m.	1.5 Days
2 Hours	8:00–10:00	10:00–12:00	12:00–2:00	6 Hours
1 Hour	8:00–9:00	9:00–10:00	10:00–11:00	3 Hours

FIGURE 8.4
Switching from daily to hourly production.

increments of time. This will help you determine a more effective labor allocation on a daily basis, leading you to create a more flexible workforce.

HOURLY PRODUCTION EXAMPLE—NABEYA

One of the companies that Miura has coached, The Nabeya Company, initially built a new machine-processing factory in 1990 at their Itonuki plant, based on the concept of a Flexible Manufacturing System. This system allowed for 24-hour continuous production by the utilization of cutting-edge automated storage systems, automated transportation, and factory automation equipment. It was a truly breakthrough factory, using the most innovative automation systems of the time.

However, after the economic bubble burst in Japan in 1991, their mass production system could not keep up with the varying customer demands. It was absolutely necessary for the company to transition from their mass production model to a high-mix, low-volume production system in order to respond to customer demand. With this vision in place, and a sense of urgency, the company launched their production transformation movement in 2002.

In order to break away from the old mass production system, which relied heavily on sales forecasts to determine production volume, the company focused on actual shipping records. They used their historical shipping records to determine the sequence of sales and base their optimal production lot sizes on this actual data. Through many trials and errors, the company succeeded in transitioning its production system to their goal of a small lot production system.

The fully automated storage system that could store and transport 1800 pallets was eliminated, and instead their work-in-process items were kept on single pallets, to more easily determine which processes have work at any given time. Before the transition, this type of information was stored in a computer system, which was accessible only to a few managers. Now, all employees can immediately see where the bottlenecks are and what needs to be done to eliminate them.

The use of computers for production management was one of the many reasons why it was so difficult for Nabeya to reduce the amount of work-in-process at their facilities. By moving to visualizing work-in-process on the shop floor, they were able to synchronize the production among different departments, little by little.

At the same time, the company also worked hard on applying Kaizen to their machinery and equipment. Before, they would attach four jobs to the same jig in a machine, creating a sizable batch. After Kaizen, they now only complete one job at a time per machine and have switched from specific automated machines to more general-purpose machines, keeping their small lot production agile and able to change at a moments notice to meet their customer's demand.

As a result of their efforts, lead times at the Itonuki plant have been reduced from 43 days down to just 4, and the amount of work-in-process reduced from 2312 jobs down to 150. Today, the company continues to develop new products and services by taking advantage of the flexibility that their production facility has achieved.

LEVELING YOUR PRODUCTION

One of the universal complaints we hear from facility managers is how to schedule production in the new system. Between juggling promise dates from sales staff to uneven labor and machinery times, every organization faces the challenge of production management. Even when production occurs smoothly according to a schedule, sometimes we suffer from not having the right products to ship to customers based on their orders. Therefore, a production management schedule must be established according to the sequence of delivery to each customer. If you don't schedule this way, you are likely to end up with a lot of unnecessary finished products and not enough correct products to fulfill your actual orders.

For instance, Kashiwa Mokko, a manufacturer of luxury furniture, implemented the Toyota Production System under the direct guidance of Taiichi Ohno, and even they struggled with production scheduling.

During one visit to the factory, Mr. Ohno pointed out that they were missing delivery dates because they were making stock that did not have a delivery date—they were making items based on a forecast. Therefore, if you make "maybe stock," which creates "stockout stock," you will never be able to locate the correct stock for delivery. The president, Mr. Michiaki

Seki, vividly remembers these times with Taiichi Ohno, telling us that no one could comprehend the true implications of what Mr. Ohno was sharing back then.

However, as the company continues to implement Kaizen, with direct coaching from Mr. Yamada now, the production department has successfully adapted to One-Piece Flow and achieved leveled production using a Pull System.

Consequently, the company transitioned its entire production system from forecasted to scheduled based on customers' orders. This transition was the first time this was achieved in the furniture industry in Japan. Mr. Seki told us during one of our tours, "I always wondered why there were missing items when we had plenty of finished product inventories. But as soon as we switched to production based on actual customer orders, eliminating finished product inventories, we had no more missing product" (Figure 8.5).

FIGURE 8.5
Mr. Seki with Mr. Ohno.

MANAGEMENT BY TIME UNITS

Consider how people train for marathons. Each runner is required to complete a long-distance run (26 miles/42.195 kilometers), and most of them have a specific finish time already set in their minds. In this situation, how are runners supposed to manage their performance, when they only have the final time in their mind? They break their run down into incremental bits, typically by each mile or kilometer, so that they can analyze that time against their goal to see if they are ahead or behind the pace they need to achieve their goal.

Runners typically have a wristwatch they check at specified intervals, so they can real-time check their actual performance against their predicted pace. These predetermined goals at each incremental distance point are equivalent in business to having hourly goals, broken out into specified cycles.

We always coach workers to manage their production process based on 1-hour increments and always take a moment to look back on each hour to make sure there are no delays. If there is a delay, workers should immediately identify the problem and formulate effective solutions in order to catch their work back up to the goal within the next 1-hour block. Production time management is also extremely effective for signaling workers whose process is ahead of schedule, so those people can be freed up to help other processes that are falling behind or spend time on improvement projects (Figure 8.6).

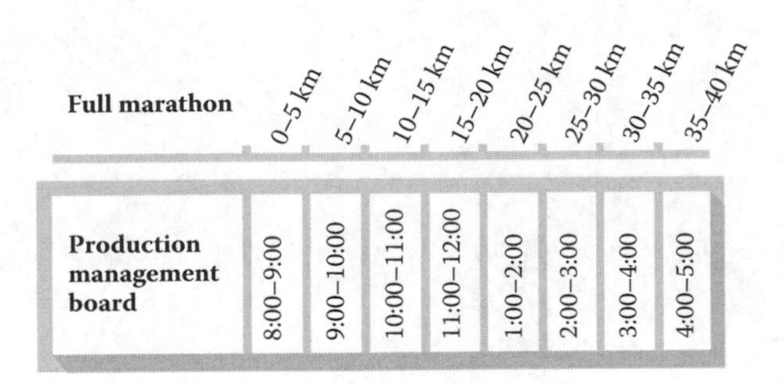

FIGURE 8.6
Marathon time increments vs. production time increments.

PRODUCTION MANAGEMENT BOARDS VS. ELECTRONIC SCOREBOARDS

When you see older images of conveyor belt production lines, you often will see an electronic scoreboard hanging from the ceiling. These scoreboards typically show production information, such as the current production goal, actual production numbers, and what the difference is between the goal and actual production (Figure 8.7).

Miura once visited an electric-appliance manufacturer to provide Kaizen coaching and suggested implementing production management boards. His suggestion was greeted with, "There is no need for production management boards. We already obtain key production progress at any time from our electronic scoreboards." While it was true you could see their production goal was 200 units and their actual production volume was 180 units (the difference between the two was also shown to be 20 units), when asked if any countermeasures had been executed to correct the situation, not one manager was able to answer.

Management boards are useless if nothing is done to solve the existing issues they highlight. The information is useful only if it is used, not just if it is accessible.

FIGURE 8.7
Electronic scoreboard example.

For example, the final assembly lines of Toyota assemble cars in a One-Piece Flow Production System, and each car flows down the line at the same rate of speed, at equally spaced intervals. Whenever there is an issue, or a specific piece of work cannot be completed within the standard work time, workers will pull an Andon cord located directly above their heads to stop the line and call for a supervisor. The supervisor will come over and, collectively, they will remedy the issue, on the spot, and bring production back to its normal state of flow.

It is management's job to ensure that each segment of work is being completed within a specified time period, and to work side by side with workers to solve issues to keep the flow of work going. Managers must sustain an effective support system among workers in order to respond to any delays in production. This is how an electronic scoreboard can be useful to the work area—as long as it can track real-time progress of each single job and helps the management to dispatch a support system to solve problems.

While electronic scoreboards can tell you when you are behind schedule, and you will hear from people who utilize them that they can respond to any abnormalities at any time by looking at the board, there is no way for the board to show them where the abnormality is taking place. As well, as long as their production lines remain behind schedule, they can never attain their ideal state.

In order to carry out all the necessary work and shipments to customers on time, you have to ensure that production is running exactly as scheduled, at every hour, to achieve your daily goals. Because of this necessity, we encourage you to begin using visual management boards as a primary tool for achieving the most important objective of your company: shipment to your customers.

UPDATE THE BOARD BY HAND

When using your visual management boards, we recommend spending the time to update the board each hour by hand. Manually writing the information connects you to the production goals and actual situation, and this allows you to quickly discern where discrepancies are taking place. Through this updating, supervisors in the business will be able to manage entire production areas and keep them on schedule (Figure 8.8).

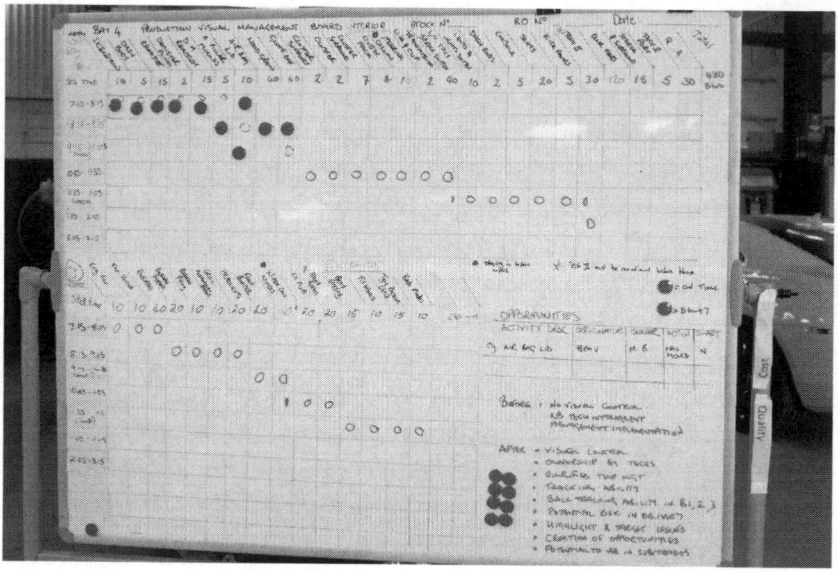

FIGURE 8.8
Example of a manually updated board.

USING VISUAL MANAGEMENT BOARDS TO REDUCE OVERTIME

A great example of using visual management boards in a nonproduction environment is Ube Kosan Central Hospital in Japan. After a number of managers attended PEC's "Train the Trainer" course, Miura went to visit the hospital to see how their Kaizen changes were progressing.

Because a patient comes into contact with multiple people during their visit to the hospital—doctors, nurses, specialists, and physical therapists—this was shared as one of the biggest challenges the hospital faced when it came to visualizing their data. Medical professionals were setting their schedules based on their own preferences and availability, leaving many patients double-booked and having to be rescheduled.

For example, say a nurse was scheduled to visit a patient to take his temperature at 9 a.m., only to discover that a physical therapist had already taken the patient away to their rehabilitation exercise and the patient wouldn't be back until 9:20 a.m. In this situation, a nurse would change the schedule for that patient on the spot and visit the next patient on their schedule to see if they were available sooner than their

Issues identified before Kaizen	
1	Waste of motion to travel to a patient room
2	Waste of time in re-adjusting schedules to execute necessary medical responsibilities
3	Waste generated from having to work overtime hours as certain operations are put off

FIGURE 8.9
Issues identified at Ube Kosan Central Hospital before Kaizen.

appointed time. Obviously, this is a waste of motion, travel, and time (Figure 8.9).

After identifying the issues in the system, their solution was to visualize the unoccupied time for each patient and create a visual management board that encompassed all of their patients. Patient names were listed on the vertical axis of the board, and increments of time, in 20-minute blocks, were located on the horizontal axis. Color-coded magnets were then used to describe exactly what service each patient was receiving at what time, as well as which medical professional was assigned to provide the patient with their treatment. This system allowed each patient's schedule, and the workload of each provider, to be clearly visible. It eliminated the issue of double-booking patient schedules and wastes of motion, travel, and time (Figure 8.10).

ELIMINATING THE WASTE OF STAGNATION BY APPLYING TIME MANAGEMENT

In order to better connect your processes together, and link them so that your production flows smoothly, you need to apply a time management system. We have already discussed ways to take your daily production down to hourly scheduling, and now we will delve further into applying time management systems to align your processes.

Let's examine how to do this using a shop floor environment where products flow through three different processes to be completed: Processing, Polishing, and Final Assembly. In the beginning, the process is being managed by daily increments.

As illustrated in Figure 8.11, Processing happens on the first day, Polishing on the second day, and Final Assembly is performed on the third day. This means that there is a 3-day lead time to final completion of an item.

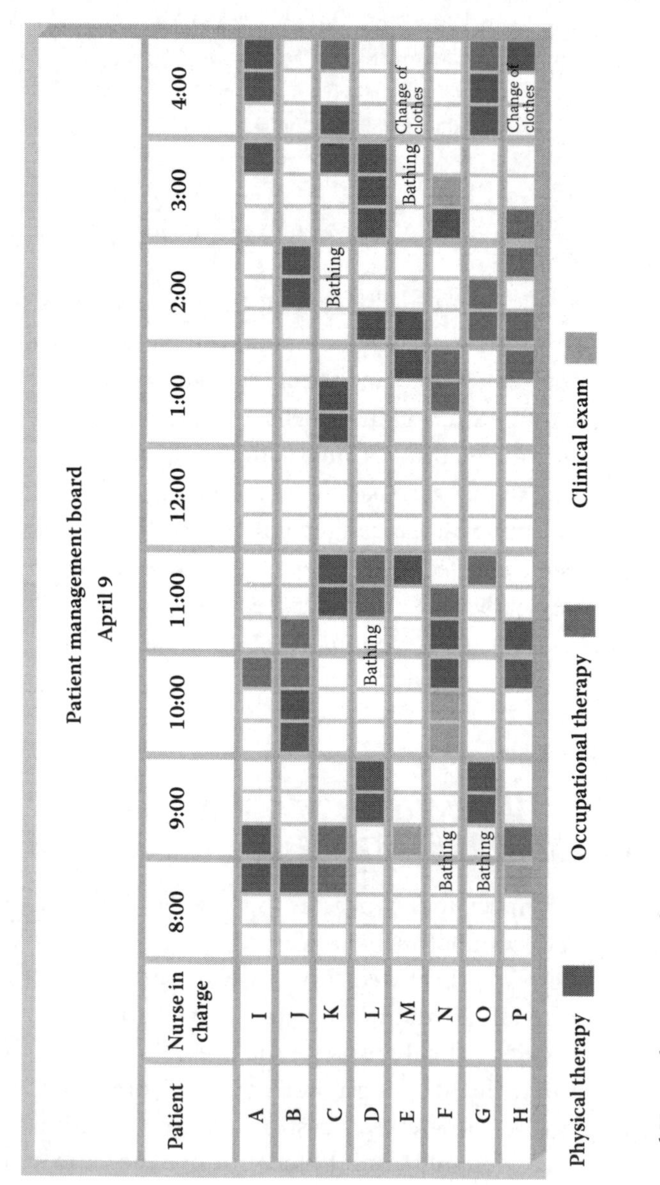

FIGURE 8.10
Ube Kosan Central Hospital patient visual management board example.

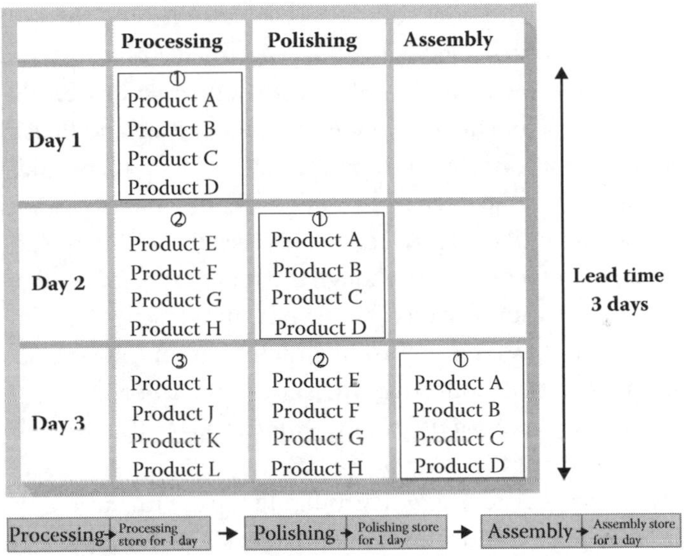

	Processing	Polishing	Assembly
Day 1	① Product A Product B Product C Product D		
Day 2	② Product E Product F Product G Product H	① Product A Product B Product C Product D	
Day 3	③ Product I Product J Product K Product L	② Product E Product F Product G Product H	① Product A Product B Product C Product D

Lead time 3 days

Processing → Processing store for 1 day → Polishing → Polishing store for 1 day → Assembly → Assembly store for 1 day

FIGURE 8.11
Daily management using Store and Fridge.

Using this method of management, only the managers for each process know when each product is going to be completed for their specific process. In this system, the upstream and downstream workers do not need to know what is going on in any of the other processes.

However, if Product A is fed into Processing at the start of the day, at 8 a.m., and is completed at 10 a.m., the product will sit for the entire rest of the day because it won't be worked on by Polishing until the second day. Naturally, each process is managed by daily increments, and each Store area for placing finished items has to be large enough to accommodate one day's worth of inventory.

In addition, when production experiences machine troubles or is missing raw materials, workers may not fix the issue right away because they think they have more than enough to complete the day's work—thus prolonging potential problems. This type of leeway does not allow a shop floor to be flexible and change to varying customer demand, as changeovers of pre-Kaizened areas typically take more than 1 hour to complete.

A management system that controls processes in daily increments makes process management easy, but it is likely to establish a culture that does not take downstream processes into consideration because, in the worker's eyes, there is no need to. Instead, when discussing improvement efforts, they will simply seek to maximize the efficiency of their own specific process.

This is the complete opposite of the very concept behind the Just-in-Time System, which signals production in each department or process to manufacture only the items needed, in the right quantity, in order to meet the delivery dates to customers. No matter how hard both Processing and Polishing work to try and satisfy the goals provided by a clear production schedule, products can't be finished until the Final Assembly process is complete. Since no revenue will be gained until final products are released to customers, such companies often run out of money, even though they have expected profits on paper.

This type of situation is called "independent efficiency," meaning each process is only interested in improving their own internal efficiency, with no regard to the overall collective efficiency of the company.

How can workers contribute toward improvement (the Kaizen culture) if they are unaware of the other processes, and ignorant to how their work affects those processes? By synchronizing upstream and downstream processes, you can tie your production facility together and, in doing so, inspire ideas for improvement from all areas of the workforce.

So how do you first begin to shift away from daily management? Start with implementing a half-day management system.

As illustrated in Figure 8.12, you can see Product A and Product B are Processed in the morning of the first day, and Polishing completes their work on the products in the afternoon of the first day. Then, the products are fed into the Final Assembly process on the morning of the second day

	Processing	**Polishing**	**Assembly**
Day 1 a.m.	① Product A Product B		
Day 1 p.m.	② Product C Product D	① Product A Product B	
Day 2 p.m.	③ Product E Product F	② Product C Product D	① Product A Product B
Day 1 p.m.	④ Product G Product H	③ Product E Product F	② Product C Product D

Lead time 1.5 days

Processing → Processing store for 1/2 day → Polishing → Polishing store for 1/2 day → Assembly → Assembly store for 1/2 day

FIGURE 8.12
Transitioning to a half-day management system.

By changing from entire day to half-day management, you can also cut your lead time in half.

and are completed by noon. In this situation, the total lead time has gone down from 3 days to just 1.5 days.

It is important to keep in mind, however, that what you are achieving with this is smaller batch sizes so your flow is faster than previously, but it is still not true flow.

Compared to a daily management system, this half-day management system is much more challenging to implement, especially in factories that often suffer machinery breakdowns. When Processing is pushed back due to machine troubles, scheduling must be readjusted because the products cannot be moved within their allotted 4-hour time period.

In this situation, management becomes so cumbersome that production becomes unable to control the situation on its own. To solve this challenge, the root causes of machine troubles and defective raw materials need to be clearly identified and addressed. Preventive maintenance on machines and an established way of controlling raw materials must be put in place, so that problems do not occur again.

After most of the causes of machine breakdowns have been eliminated, the store area can be reduced for each process, from 1 day down to a half-day of inventory, saving you space and money (Figure 8.13).

	Processing	**Polishing**	**Assembly**
Day 1 a.m.	① Product A		
Day 1 p.m.	② Product C	① Product A	
Day 2 a.m.	③ Product E	② Product C	① Product A
Day 1 p.m.	④ Product G	③ Product E	② Product C

Lead time 1.5 days

FIGURE 8.13
Store area reduction for half-day management.

After reaching this level of management, it will become impossible to increase the productivity efficiency any higher in a given process on its own, and it will become necessary for both preceding and proceeding processes to start sharing key information.

For example, say that Final Assembly will have to stop for the afternoon if they cannot confirm a delivery of all necessary items to be assembled in their work area by 1 p.m. In this case, the leader of the Final Assembly process begins visiting his upstream process (Polishing) to make sure that the lots (Product A, C, E, and G) will be completed in time for the Final Assembly process to receive the items and finish their work for the afternoon. Thus, the sharing of vital information happens and schedules can be adjusted accordingly (Figure 8.14).

There are many reasons that schedules must be changed, such as a customer changing a delivery time or cancelling their order, or the Processing department could be finding many defects and needing to stop their work to investigate why. In these situations, production planning for the Polishing process should be adjusted so the sequence flow of products remains undisturbed, and the Final Assembly process should also adjust its schedule accordingly with the new upstream changes so that stoppages in production can be avoided.

Implementing a time management system helps people visualize workflow and potential problems. It greatly strengthens the capacity and flexibility of our workplaces, allowing people to solve problems one by one. In addition, in a case where the items that are to be assembled need to be

	a.m.				p.m.			
	8:00	9:00	10:00	11:00	12:00	1:00	2:00	3:00
Product A	Production Kanban A Production Kanban A	Production Kanban A	Production Kanban A Production Kanban A	Production Kanban A	Production Kanban A Production Kanban A	Production Kanban A	Production Kanban A Production Kanban A	Production Kanban A
Product B	Production Kanban B	Production Kanban B		Production Kanban B Production Kanban B		Production Kanban B	Production Kanban B	Production Kanban B
Product C		Production Kanban C	Production Kanban C		Production Kanban C	Production Kanban C		Production Kanban C

FIGURE 8.14
Example of product flow using Kanban.

completed and shipped out by the end of the same day, any workers in the shipping department can view the Production Management Board and quickly learn exactly where items are complete or behind schedule.

One of the true benefits of a time management system is that everyone in different processes will begin sharing information, determining true priorities for production, and creating stronger teamwork between all processes.

Consequently, each workplace begins to contribute to creating a more desirable organizational culture by striving to achieve a total efficiency flow, instead of independent process or departmental efficiency.

After a whole-organizational flow culture has been established, it is time to introduce the 2-Hour Time Management System (Figure 8.15).

In the 2-Hour Time Management System, Product A is scheduled for Processing between 8 a.m. and 10 a.m., Polishing between 10 a.m. and 12 p.m., and Assembly between 1 p.m. and 3 p.m. A total lead time in this scenario is 7 hours (2 hours for each process, multiplied by three processes, including a 1-hour lunch break).

A daily management system required a lead time of 3 days; however, this 2-hour system cuts the lead time down to less than 1 day, which allows for same-day shipping.

This also decreases the number of work-in-process items to only an amount which can be completed in 2 hours. Work Kaizen will continue

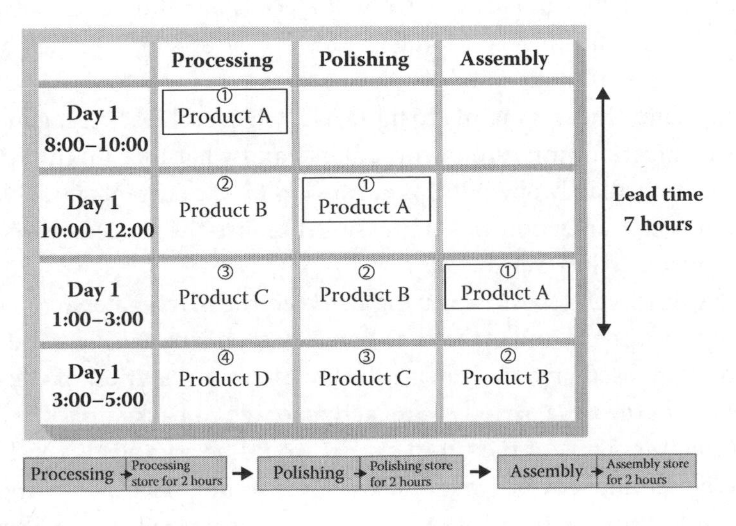

FIGURE 8.15
2-Hour Time Management System.

to increase from workers, as you will need to develop faster turnaround times for your processes.

Naturally, when processes are brought closer together, it creates better teamwork and communication. Production becomes able to operate in a One-Piece Flow mentality through vigorously eliminating the wastes of motion, transportation, and stagnation. Any handling of jobs across processes can be managed by intervals of a certain time period and wasteful administrative tasks can be eliminated.

PUTTING PRODUCTION BOARDS INTO ACTION—TANAKA FOODS

Tanaka Foods is a well-known manufacturer of what is known as "furikake," a dried seasoning used to sprinkle on rice, in Japan. Tanaka Foods initially began with only one machine to mix flavored raw materials, but they had 11 packaging machines. With this equipment limitation, the mixing process would batch-process 2 days worth of inventory in order to satisfy the capacity of the packaging process. To help them streamline their production and become more profitable, they brought in Miura to help them implement Kaizen in their facility. Since they have begun implementing Kaizen practices, they have been able to get to just-in-time production of particular raw materials based on each packaging machine's specific finished product needs.

The first step Tanaka Foods took to achieve this was to draw out their current state. This was mainly to figure out when each packaging machine required a certain mixture of ingredients and what that mixture was. It became immediately obvious upon drawing up the current state that production synchronization was impossible because the processes were currently unable to be leveled.

To fix this issue of synchronizing between the Mixing department and the Packaging department, Takana Foods began using a visual production board. They used arrows drawn into the afternoon section of the board to represent the time periods that each machine in Packaging would run, including the name of the product that would be packaged. The Mixing process then knew which materials needed to be mixed in the morning to provide Packaging with the correct product needed in the afternoon (Figure 8.16).

Schedule for mixing process in p.m.	Packaging machines										
	Machines 1	Machines 2	Machines 3	Machines 4	Machines 5	Machines 6	Machines 7	Machines 8	Machines 9	Machines 10	Machines 11
8:00	A	B	D	F							
9:00					H K		M	N	P		
10:00					I			O		RS	UV
11:00		C	E	G	J	L		Q		T	
12:00											
1:00	A	B	D	F	H	K	M	N	P	R	U
2:00	↓	↓	↓	↓	I	↓	↓	O	↓	S	V
3:00	↓	C	E	↓	J	↓	↓	↓	Q	↓	↓
4:00	↓	↓	↓	G	↓	L	↓	↓	↓	T	↓

FIGURE 8.16
Tanaka's mixing process production management board.

For example, Packaging Machine 1 is scheduled to pack Product A between 2 p.m. and 5 p.m.; therefore, the mixing of Product A needs to be scheduled between 8 a.m. and 9 a.m. In this particular situation, there is a stagnation time between Mixing and Packaging of 4 hours.

After their initial Kaizen implementation, the amount of already mixed work-in-process materials was reduced from 2 days down to half a day. Work-in-process items were reduced to a quarter of their previous amount, and the lead time from the start of Mixing to the finishing point of Packaging was reduced to 9 hours, which allowed for rush orders that were received by 8 a.m. to be shipped out the same day to their customers.

Understanding precisely what materials are needed, in what quantities, and at what time is absolutely essential in order to implement a production system that can be flexible enough to meet your customer demand. If this understanding is achieved, Production will soon be able to identify any bottleneck processes and promote Kaizen that targets those specific bottlenecks. Production will also be able to pull in design and purchasing divisions to help them streamline production.

REDUCING PRODUCTION BREAKDOWNS

It can often be difficult to see all the hard work that has gone into planning for smooth production, and it is often only when things go awry that people begin to appreciate the difference between the current improved flow and the previous, un-Kaizened workflow. Breakdowns are often the cause of production going awry, and it is important to eliminate these breakdowns so that all your hard work to create smoother production can be maintained.

Potential breakdowns in production can easily be classified into two categories: internal causes and external causes.

Internal causes can include the following:

- Staffing challenges: Either an inadequate number of personnel or lack of skills
- Equipment availability: Capacity fluctuation due to maintenance issues or breakdowns

- Erratic quality: Uneven processing or quality of inputs which change the quality of the final product
- Problems in production scheduling and flow

All of these factors are within the control of the organization. If you believe they cannot be fixed within your facility, then you need to challenge your own assumptions.

External causes that can stop your production include the following:

- A drastic shift in customer demand
- Technological changes and advances
- Reliability of supplied material, components, and parts

While these are important causes that can have an effect on your processes, you should always focus on the issues that you can address. Look through the list of internal causes again. How can you apply Kaizen to each of those factors to ensure they never again stop your production?

EQUIPMENT MAINTENANCE

In factories in Japan, maintenance is not carried out by dedicated engineers who have received special training, in the same way that cleaning isn't carried out by a special crew of people. Instead, the machine operators, who have been trained, perform routine maintenance and repairs (Figure 8.17). Because they use their machines on a daily basis and know them inside and out, they are the first to realize when maintenance or repairs are needed. It could be said that this mindset has been inherited over hundreds of years in response to the craftsman mentality that says, "The tools you use must be polished with your own hands."

Another important thing to note is that workers do not wait to perform maintenance until after a machine has broken down. Instead, workers perform maintenance on each machine autonomously and check on the machines capabilities daily to proactively eliminate any chance of machine breakdowns due to mechanical abnormalities.

The first step in implementing this style of maintenance is to decide on daily checkpoints for each machine and have workers begin to carry out a maintenance inspection daily, before each shift begins. Separate lists of

FIGURE 8.17
Equipment maintenance by a machine operator.

maintenance and inspection checkpoints also need to be determined to direct how maintenance will be performed at daily, weekly, monthly, and yearly frequencies. This needs to be decided on in order to maintain a high level of quality and safety standards for overall production.

AN EXAMPLE OF DAILY MAINTENANCE IN PRACTICE

At Kashiwa's Furukawa Factory in Japan, the employee in charge of each process performs his or her machine inspections daily, before work begins. If any abnormalities are discovered, the responsible worker fixes the problem and documents their countermeasure on a daily inspection sheet.

Each month, the production line leader checks each machine's daily inspection sheet for any abnormalities encountered over the month. If necessary, they will add additional inspection criteria for the upcoming month. This new checklist is only added to help prevent the same errors from occurring in the future (Figure 8.18).

Yearly checklists are added to, or tasks can be omitted from them, depending on historical data. These are adjusted, the same as the monthly checks, to ensure the same errors do not happen on a recurring basis.

Machine name: Running saw machine						Inspection for August
No.	**Inspection item**	**Inspection method**	**Judgment criteria**	**8/1**	**8/31**	**Solutions for abnormalities**
1	Oil level	Is there enough oil level in the cup?	More than mix level	O	O	Filled up oil to the max level
2	Air pressure	Check air pressure gauge	5–6 kgf/cm^2	O	O	Reported to production leader and followed his instructions
3	Belt tension	Check to see if tension is appropriate	No loosening over 5 mm	O	O	Readjusted tension of the belt
					O	Replace the belt itself

Status legends:

OK: O NG: X (Not a breakdown, but treated to standards) Example: Oil was checked and filled up even though it was not below the min. level.

After treated: ● (Identified as a breakdown and treated to normal) Example: Belt was replaced as it had cracks.

FIGURE 8.18
Maintenance inspection sheet.

It is important to have the people who work on machines trained and responsible for the maintenance of their machinery because they know the machine best. When only qualified engineers from a specialized division are allowed to perform maintenance, you run the possibility of a serious threat to production happening, as there is a much higher probability that an urgent abnormality will not, and cannot, be treated soon enough. When the workers who operate machines are trained, they can immediately respond when they hear something strange in how the machine is running, and they can detect issues themselves instead of having to wait for someone to become available for diagnostic work.

Because the person doing the work can detect issues, and appropriate countermeasures can be applied immediately, you can minimize the impact abnormalities have on your production and often avoid machinery breakdowns. This is of vital importance to achieving smooth and consistent production, so we cannot stress enough the importance of educating your workers so they are able to perform daily maintenance inspection.

MATERIAL CHECKING

Material checking is a vital part of your daily control to ensure you never run out of materials; however, this is not something that should be conducted every morning by the plant manager. Material checking must be practiced by the visualization of material storage places. By asking yourself a series of questions, you can examine and check your materials (Figure 8.19).

First, ask yourself, is the lead time from the placement of an order for materials to the delivery of those materials understood? As long as the lead time from order to delivery is understood, stock on hand can be strictly controlled.

Next, is the volume of component materials on-hand understood? Any excess amounts of material in the system to compensate for factors such as unreliable suppliers, delays in logistics, equipment breakdowns, or any other anticipated occurrences are a waste. In order to eliminate this excess, the reduction of inventory needs to come from better inventory control, logistics, and overall system reliability.

To reduce the overall volume of stock on hand, consider the best way to visualize the stock so that everyone can see how much is needed and how much is waste. As well, begin to implement Kaizen for your logistics, and

FIGURE 8.19
Example of material checking at a furniture manufacturer.

practice preventative maintenance on equipment so that you don't lose time due to mechanical breakdowns.

A few more questions to answer are the following: Are materials subject to quality checks? Are materials properly stored?

Among many types of materials, changes and variations in quality are expected to occur as time passes, so quality checks must occur frequently to spot problems before they occur. As well, the storage location and method for each type of material must be observed and controlled closely to adapt to changes in quantity and storage needs.

THE FIRST-IN, FIRST-OUT METHOD

In order to create material ordering and receiving systems, visualized methods must be established to quickly and easily understand the status of materials at the time of order placement.

In one location that Collin coached, parts in need of refurbishing would be received from customers, assessed for the nature of their repairs, and quoted for the work that needed to be done. The pieces were then placed on the shop floor, waiting for approval from the customer. You can probably guess at the issues that this particular company faced with this method of storage while waiting for approval. Some workstations were overloaded with pieces on which they could not begin work, which got in the way of work that could be done and created a huge waste of Stagnation.

A real effort should be made to place only the work that is ready to be processed on the shop floor, and a continuous flow with all parts available to complete the work should be strived for. If something is not ready to be worked on, it needs to be stored elsewhere, not on the shop floor, so that items get worked on in the order they are received onto the shop floor, not into the system in general. This creates a first-in, first-out (FIFO) system and allows work to flow appropriately. Therefore, any reduction of inventory must come from better control (Figure 8.20).

HAVING THE RIGHT MATERIALS

Engineering designs and specifications change as time passes—this is the nature of work. Therefore, it is important to have only the materials that

FIGURE 8.20
Example of a FIFO system in practice.

FIGURE 8.21
Utilizing Kanban for inventory control.

are truly necessary on hand and not overstock or keep items that are no longer necessary due to design changes. Inventory levels must constantly be assessed against current production requirements.

It is essential to monitor inventory on the basis of inventory turnover. By classifying on the bases of how often inventory turns, slow moving inventory can be quickly identified, as well as which items are special requests from certain customers, which often proves to be a real challenge. Because the real issue is turnaround time, a real reduction in inventory can be accomplished by improving flow time through operations. Quick turnaround time also minimizes the need for on-hand inventory of special customer products (Figure 8.21).

CONCLUSION

Using techniques such as visual management, and the various tools to help you achieve visual management, you can balance your workload and begin to see the true lead times of your products, from the office through shipping.

Scheduling your production in incrementally smaller periods of time helps you synchronize your processes and bring together your teams. By implementing an effective maintenance routine, including training workers to maintain the machines they use, you will save time and prevent production breakdowns.

9

Waste Elimination in Practice

After synchronizing your production processes and managing in smaller and smaller increments, you will once again be able to visualize waste in your facility. Just as True Kaizen is a never-ending cycle because you are always uncovering new things and learning more, you will find that there is a never ending supply of waste to be eliminated at your facility. As soon as you fix one situation, the efficiencies in that area will make other wastes apparent. Thus, we have decided to dedicate another chapter to further explore the practice of waste elimination.

A very important way to determine what is actually waste is to make sure you know who your real customer is. Often, we find ourselves performing what we think are value-added tasks, yet when we meet with the customer and view it from their perspective, we find they do not see these tasks as valuable at all. Therefore, you must first understand how your customer is using your products, so you can adjust your production to meet their needs.

As an example of adjusting to your customer's needs, let's take a look at a filmstrip manufacturer in Japan.

This film manufacturer once produced a filmstrip that shipped at a default length of 4060 mm. Company representatives wanted to gain better insight into their customers, so they went and visited a number of customers and discovered that they were cutting the film into either lengths of 1020 mm or 2520 mm. In doing this, they were ending up with an excess of 1120 mm length of film. Even though the company was running its production based on customer demand for their product, they did not understand the actual physical needs of the customer—they were basing everything on the 4060-mm length they chose.

By connecting with their customers, the company discovered that the customers could actually yield multiple filmstrips of 1020 mm from a single filmstrip purchased. In order to more quickly meet their customer's true demand, they cut down their product and shipped in specifications of 2040 mm and 2520 mm, instead of their original 4060 mm. In doing so, they adjusted their production schedules so that the products were delivered to each customer exactly when the customer used the product. This allowed the company to completely eliminate the need for transporting and storing finished products in the production facility. As a result, the company was able to reduce their inventory by 1800 tons, saving them over $6 million, as well as becoming better able to serve their customer's needs.

In another case, a chocolate manufacturer, Tirol-Choco Company, has been producing the best-selling chocolate product in the industry in Japan, called "Tirol Choco." The company has never offered any discounted pricing for their products. Despite their confidence in their own products, and their proven strength in new product development and marketing, one of the challenges that Tirol-Choco was facing was the management of leftover packaging materials.

To solve this challenge, the sales director started communicating with the production division on a daily basis and made sure that each item type could be sustained until the respective packaging materials were used up completely. Generally speaking, sales divisions tend to want to have more finished product inventory than they are currently able to sell to customers in order to satisfy any potential sales opportunities. This company is a rare case in this respect because the sales director keeps a close eye on the inventory level of packaging material and commits his sales division to continue selling the same products until the inventory expires.

Before this effort, the company made a rule to keep approximately 2 weeks of packaging material for the regular products. The company started letting their suppliers know that their production volume needed to be achieved in 3 days and adjusted the ordering point of packaging supplies so their suppliers could deliver exactly what is needed the day before the products were finished. As a result, the inventory amount of packaging supplies was drastically reduced, from 2 weeks to only 3 days, and they have kept it sustained at this level.

A FLOW MENTALITY

Once you start implementing a just-in-time system, you will begin eliminating waste all over your facility because it is simply no longer necessary.

A high-end furniture company that Miura coaches, Hida Sangyo, has been providing exceptionally high-quality wood furniture by using the Japanese traditional craftsman skills for over 90 years. The company has switched its production system to a build-to-customer-order system over the last few years. However, during the beginning, they were still relying on mass production for furniture parts, such as table and chair legs, which require multiple phases of machine processing. Because of this, the processing part of production planning was determined based on the forecasted sales, and then the parts were being stored internally on the production floor. Around this time, the factory held work-in-process items for an average of 12.5 days.

After the factory received customers' orders, appropriate work-in-process items were taken from the shelves and fed into the production lines to be painted and assembled into final products.

To start changing their approach to production, they began feeding one customer order at a time into the very first process, in order to establish One-Piece Flow production. In doing so, the amount of work-in-process was significantly reduced, which created a much stronger sense of teamwork between the processes, and they did away with the mass production of individual parts, removing much of the waste of motion in their facility.

Mr. Takayuki Makimoto, a section manager at Hida Sangyo, told Miura that, as a result of this transition, the amount of work-in-process steadily decreased, and because the amount of necessary workers in the area dropped from 23 to 19, they were able to create four flexible workers to move about the facility and assist with various processes as needed (Figure 9.1).

Furthermore, they used to require three workers who were solely responsible for transporting packaged finished furniture to the shipping area and rearranging them on the loading dock based on the pick-up schedules of the respective outbound logistic companies. Currently, this rearranging is being done in the packaging process within the production line, and each finished item is being transported to the shipping dock based on the pick-up schedules.

This is possible thanks to the transparent communication of their visual management system. As a result, only two workers are now needed to manage the entire supply chain of finished products, and one worker has become a flexible worker and handles other important tasks for the company (Figure 9.2).

Once the waste of Stagnation is eliminated, you will be able to remove the waste of Motion as well, which leads to increased productivity.

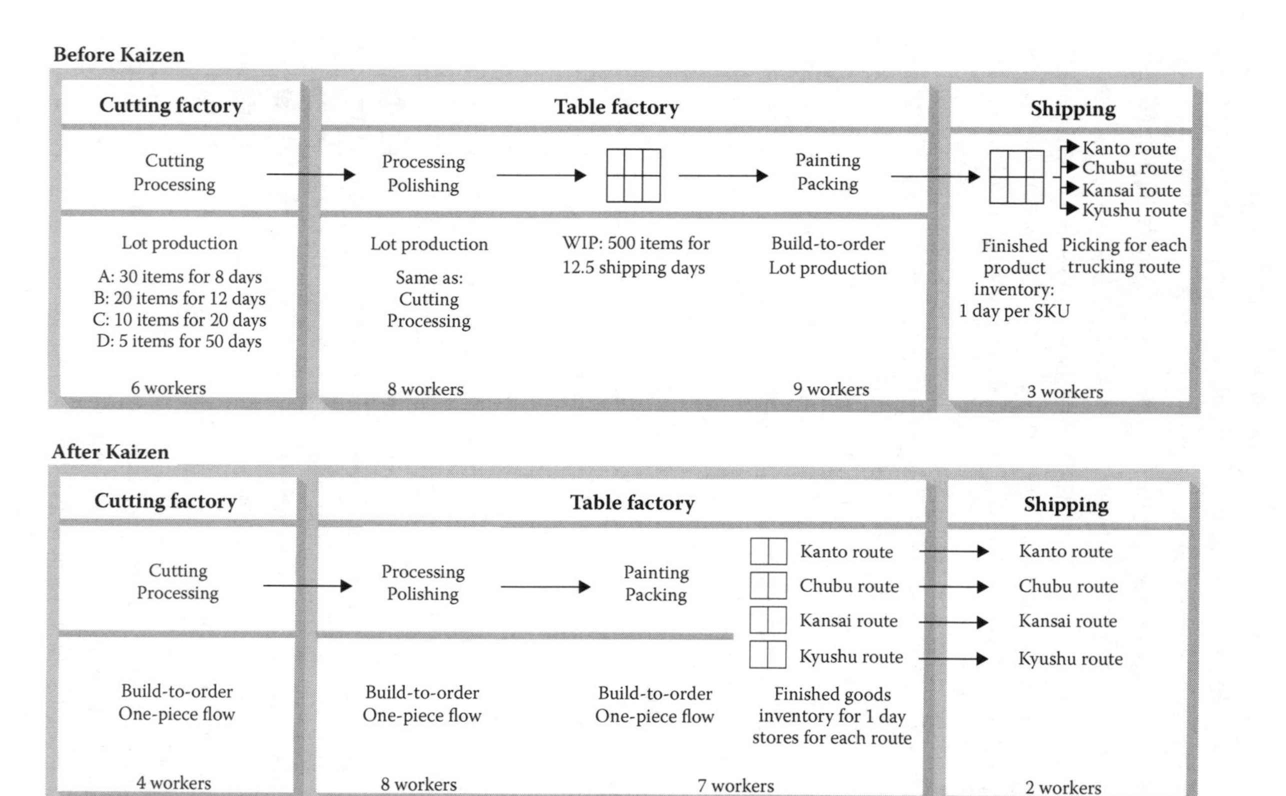

FIGURE 9.1
Before and after Kaizen at Hida Sangyo.

Packaging machines			
	Before Kaizen	After Kaizen	Effectiveness
Lead time	20.1 days	5.2 days	Reduction of 14.9 days
WIP	26 million yen or $261,000	4.6 million yen or $46,000	Saving of 22 million yen or $215,000
Number of workers	23 workers	16 workers	5 flexible workers

FIGURE 9.2
Communication of before and after Kaizen at Iida Sangyo.

DEEPER EXAMINATION OF THE WASTE OF MOTION

Once you have your facility working smoothly, and products are moving according to One-Piece Flow, you can begin focusing on the waste that occurs in the individual tasks that make up each process. Here is where you will find the waste of Motion.

The waste of Motion pertains to any wasteful movement a person makes when completing a specific task. The waste of Motion can be further broken down and classified into each respective part of the body, and the movements made with each, such as bending, lifting, and searching (Figure 9.3).

When assessing an area, pay attention to how many steps people are taking to travel to their destinations. Most of the small tasks within a particular work area require a few steps, but if you focus on all the steps they're

Waste of Motion	
Legs	Reduce the number of footsteps
Hands	Reduce the number of hand movements
Turning	Eliminate turning of the body
Bending	Eliminate work to be performed below your waist
Eyes	Manage by fixed-placement and labeling of items so that searching is no longer necessary

FIGURE 9.3
Waste of Motion—defined.

Waste of Motion	
Steps	One step = 0.8 seconds
Turning	90 degrees = 0.6 seconds
Hand motion	20 cm = 1 second

FIGURE 9.4
PEC's quantification of Motion waste elimination on time savings.

taking, you can begin to move everything closer together, reducing the amount of necessary steps.

After eliminating the waste of Motion, no matter how small, it is imperative that you measure the results of the changes you've made. PEC uses certain metrics to quantify each Kaizen for its effectiveness.

To quantify Motion waste elimination, PEC equates removing one step to saving 0.8 seconds, removing a 90 degree turn of a person's body as saving 0.6 seconds, and eliminating a hand motion for 20 cm (7.9 inches) as saving 1 second (Figure 9.4).

Providing these types of guidelines for quantifying Kaizen results allows everyone to quickly understand and gauge the effectiveness of each Kaizen implementation as well as their personal contribution to the larger company goal. After eliminating a number of steps, hand motions, and turning, start bringing everything closer together (a concept called "Majime") to further eliminate the waste of Motion.

Removing Wasted Motion while Serving Customers—Suzette Co.

A western confectionary manufacturer, Suzette Co., has been applying Kaizen to develop a more effective way of interacting with customers in their retail stores, which are located inside many popular department stores across Japan. Mr. Masakazu Takeuchi, the retail manager of their Osaka store, one of their busiest retail branches, closely observed the customer interactions on site for months to understand the process, from choosing the product through the final payment process. He focused on looking for the wastes of Motion and Stagnation in the movement of the employees serving the customers.

This is the process flow of a common interaction that Suzette employees have with customers, which Mr. Takeuchi, the retail manager, observed:

- Receiving an order
- Picking appropriate fresh cake(s), as ordered by the customer
- Confirmation—making sure the right products were picked
- Placing the cake(s) into a box
- Placing refrigerant into the box
- Handing the product to the customer
- Receiving payment
- Handing the change (if any) and receipt to the customer

What he found most problematic in the cycle of customer interactions was a long distance between the place where the cake was placed in the box and the place where the necessary refrigerant was kept. A worker would have to walk six steps to the refrigerator—a process that was reduced to only two steps by bringing the storage closer to where the cakes were placed in their boxes.

Another improvement that was made based on Mr. Takeuchi's observations was the location of the tape and company stickers used to seal and package the boxes. This was brought closer to where the product was set in the box, shortening the total packaging time by 12 seconds.

The time required to complete one cycle of customer interaction, from receiving an order to handing the receipt to the customer, was reduced from 3 minutes to 2 minutes 48 seconds. This resulted in a 7.2% increase in the efficiency for this particular job (Figure 9.5). When asked if he was satisfied with the numbers he had achieved, Mr. Takeuchi replied, "I am happy that the retail staff's response time to each customer has dramatically sped up, which reduces the waiting time for other customers. I will continue to keep removing small wastes on a daily basis to maximize the level of customer satisfaction."

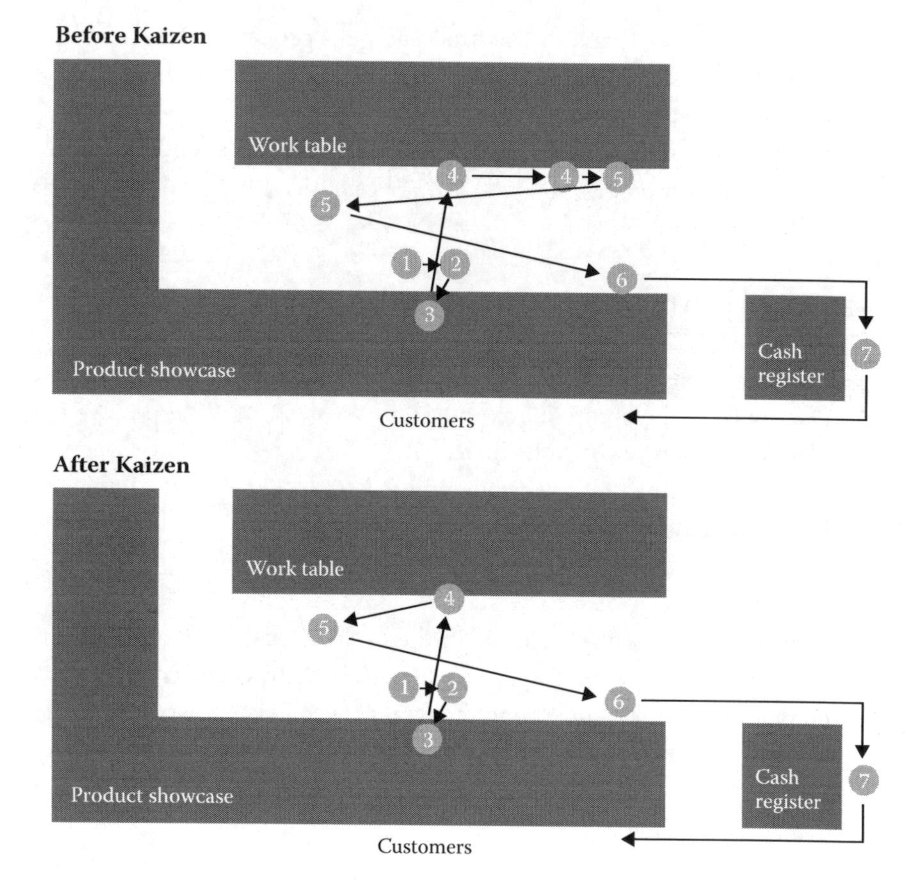

FIGURE 9.5
Before and after Kaizen at Suzette Co. retail store.

EVERY MOTION MATTERS

Furukawa MFG/OLD Rivers, a manufacturer of automated packaging machines was focused on removing Motion waste in their final inspection process for their vacuum packaging machine line. The process was currently taking 2 days (16 hours) to inspect five vacuum packaging units. Their cycle time to inspect one unit, at that time, was 3.2 hours (Figure 9.6).

To help remove wasted motion, they observed the process. It was found that the final assembled vacuum packaging machines, which were about 50 cm (20 inches) tall, were transported on pallets into the final inspection area. The quality inspectors performed most of their checks below their waist, causing them to have to bend down a number of times per unit. All

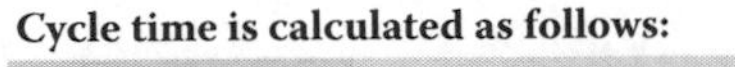

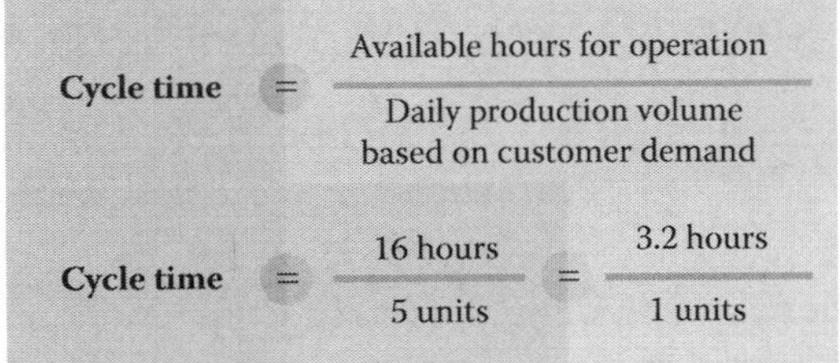

FIGURE 9.6
Vacuum packaging machine inspection cycle time.

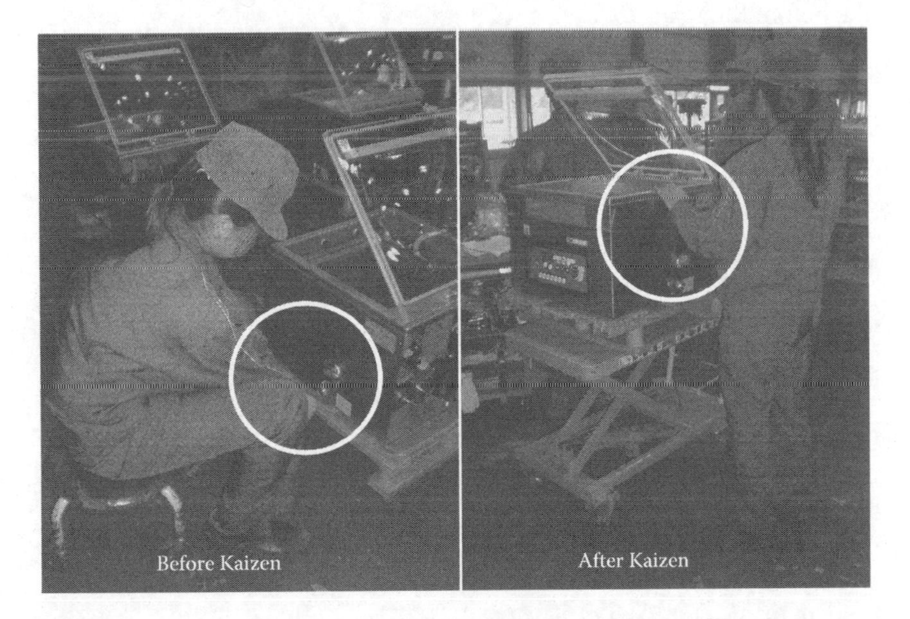

FIGURE 9.7
Before and after Kaizen—tool and work cart modifications.

the inspectors agreed that it would be much easier if the units were higher up, but since the units were transported on pallets, they were not able to come up with a solution to position the units higher off the floor at that time.

Their eventual solution was to keep finished units on the work carts, and they changed the floor layout so they could perform all the necessary inspections without needing to bend over. They also had to modify the tool carts so that the tools became easily accessible and above their waistline. After this change, it only took 1.5 hours to complete the final inspection for a single unit, cutting their process time in half* (Figure 9.7).

* Hitoshi Yamada, *Forging a Kaizen Culture*, Enna, 2011.

REMOVING MOTION WASTE IN ADMINISTRATIVE TASKS

We often picture Motion waste as something that is unique to tasks that involve a great deal of physical labor, but there is actually Motion waste that can be eliminated in a wide variety of operations.

A group leader from the NJR Fukuoka Company, a manufacturer of industrial semiconductors, focused on simplifying the data entry tasks required for each computer terminal, which were located in every production process within the factory. Below is what the production flow looked like before Kaizen:

1. Items transported from a fridge area to the terminal
2. Recording of the incoming item through the terminal
3. Items transported to the inspection machine
4. Processing or inspection takes place
5. Items transported to the tester
6. Testing takes place
7. Items transported to the terminal
8. Completion of incoming data entry
9. Items transported to a store area

To change their process, they resequenced the order in which the terminal was used and where inspections and testing physically took place. Here is how their process changed:

1. Items transported from fridge area to terminal
2. Inspection takes place
3. Items transported to tester
4. Testing takes place
5. Items transported to the terminal
6. Recording of incoming items through the terminal
7. Completion of incoming data entry
8. Items transported to a store area

As a result, they discovered that the initial data entry process and the final data entry process were actually repeated processes. Based on this finding, their existing terminal system, used for the final inspection stage, was altered so that the initial data entry and the final data entry processes

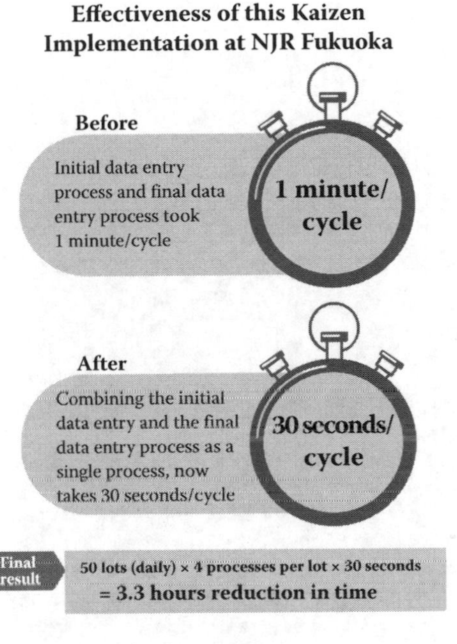

FIGURE 9.8
Effectiveness of Kaizen implementation at NJR Fukuoka.

could be done as a single data entry process. Unification of these processes took workers from 1 minute of data entry down to 30 seconds: a savings of 50%.

To view this in a larger context: the production floor was producing 50 lots across four different processes on a daily basis, and with this combining of data entry processes, they were able to reduce their overall daily processing time by 200 minutes (Figure 9.8).

FLEXIBLE WORKERS AND THE WASTE OF SPACE

A manufacturer of fried snacks in Japan, Maruka Foods Corporation, was using automated packaging machines to wrap finished products individually. Finished products were then boxed up and temporarily stored on site. Then, these boxes were transported to the final packaging process area and each individually wrapped product was removed from the transportation box, inspected, and reboxed back up for final shipping.

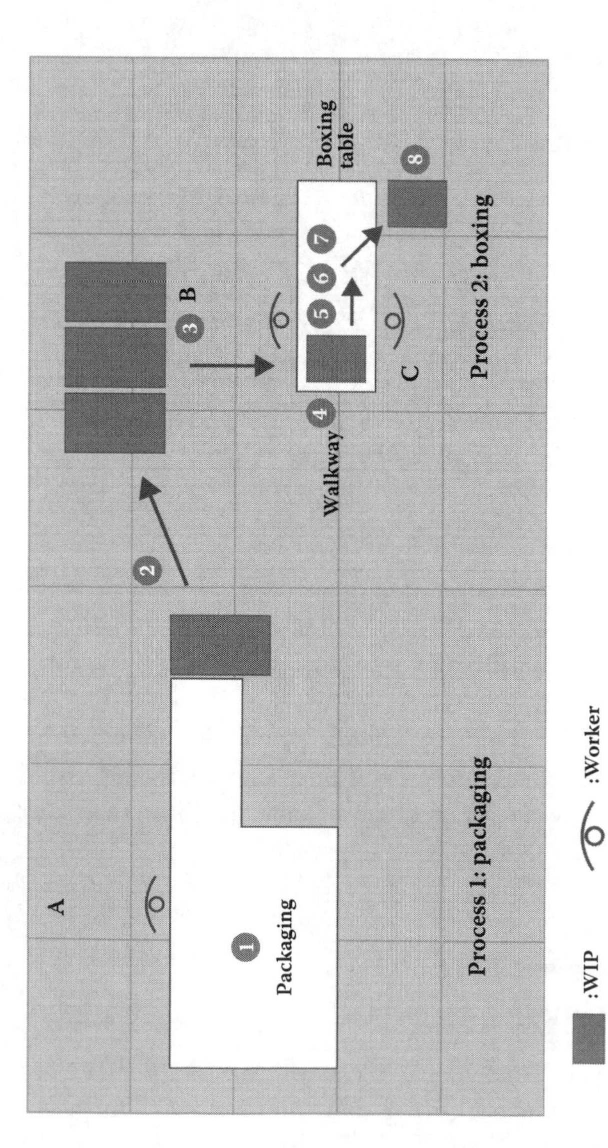

FIGURE 9.9
Maruka factory layout before Kaizen.

There were a total of three workers to complete this process before they began their Kaizen implementation. One was assigned to operate the automated wrapping machine, and two were assigned to transportation and final packaging of the products.

Looking at the pre-Kaizen process in Figures 9.9 and 9.10, the processes that could be considered value added for the customer were packaging (Process 1), sorting (Process 7), and boxing (Process 8).

The non-value-added, but necessary, task was Process 6 (checking expiration dates). All other processes were absolutely unnecessary—textbook examples of waste. Unfortunately, these processes were necessary to perform at the time because the value added processes were located so far away from each other.

To fix this, the checking of expiration dates and final boxing were done right beside the automated wrapping machine. As a result, they were able to cut their necessary processes in half—from eight down to four. This cut out the non-value-added processes: transportation (Processes 2 and 3), taking products out of the box (Process 4), and turning over the product (Process 5).

Since the processes were now connected closer together, the person who ran the automated wrapping machine could use his idle time to help other

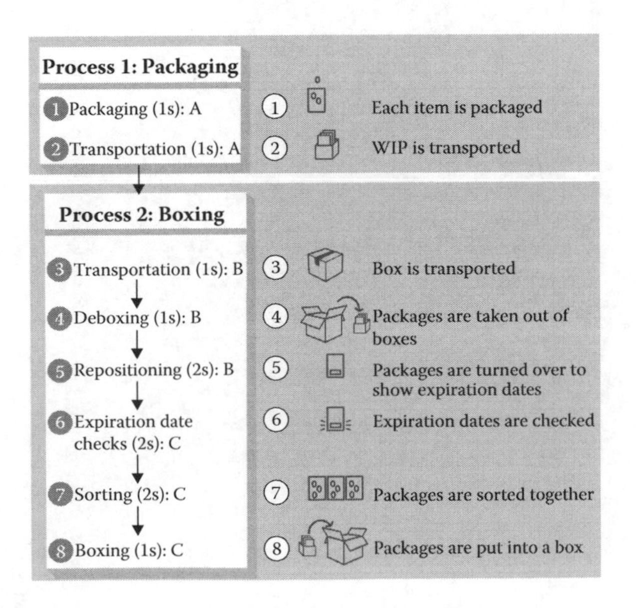

FIGURE 9.10
Maruka production process flow before Kaizen.

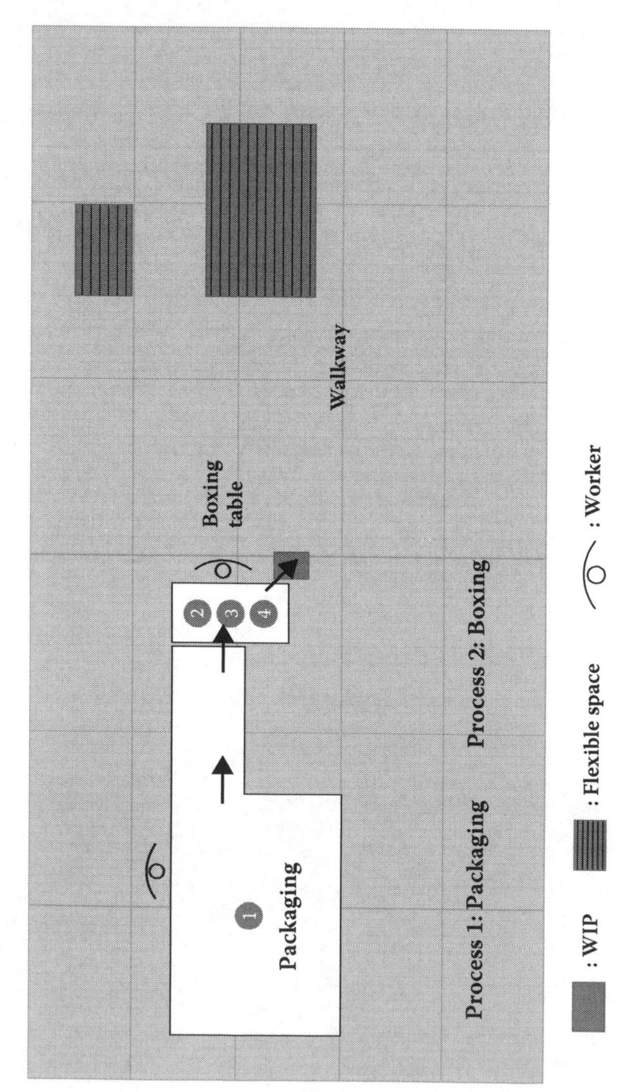

FIGURE 9.11
Maruka factory layout after Kaizen.

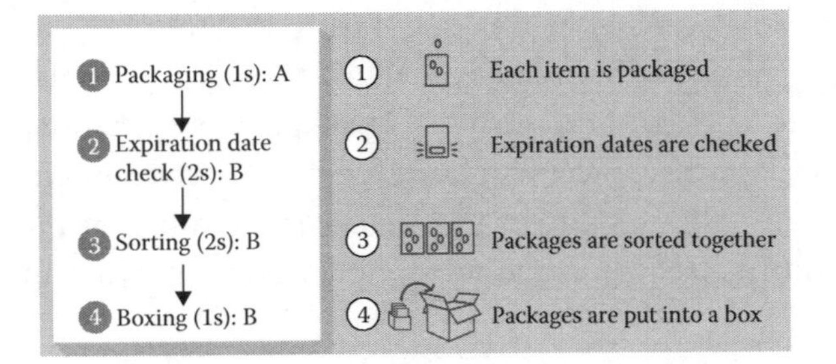

FIGURE 9.12
Maruka production process flow after Kaizen.

workers in both the packaging and boxing processes. The number of necessary workers for these processes was reduced from three workers to one worker, creating two flexible workers who could be reassigned to other important tasks for the company (Figures 9.11 and 9.12).

ADDRESSING TRANSPORTATION WASTE WITHIN YOUR FACILITY

The primary cause of Transportation waste is that items are placed far away from where they are needed. Processing is separate from final assembly, and often, many facilities don't even store the raw materials they need on location, necessitating huge amounts of Transportation waste.

Just like the waste of Motion, the waste of Transportation pertains to how humans move at any given time. What separates the waste of Transportation from the waste of Motion is that Transportation always involves some use of delivery equipment. Any work that is dependent on the use of trucks, forklifts, manual lifts, transportation carts, and conveyor belts does not add any additional value to the end product from the customer's point of view.

When we transport materials from one process to another, we often go through a number of unloading and reloading tasks along the way. For example, if raw material is delivered to a factory and stored in the receiving department, it will later be loaded onto a transportation cart and transported to the raw material storage area. From there, a certain amount of raw material is reloaded onto another cart and is fed into the appropriate production line. You can see how the waste begins to pile up! And that is without even counting the time spent searching for the right material or reloading if the wrong item was grabbed.

In order to eliminate the waste of Transportation effectively, the best thing to do is bring processes closer together: a concept we have mentioned before, called "Majime." By streamlining different processes into a single continuous flow, you eliminate the need to transport materials in the first place.

MAJIME:
The process of bringing items closer together.

The following are a few examples of companies we have worked with and how they have restructured the layout of their organizations to eliminate the waste of Transportation.

Moving Closer Together—Hida Sangyo

Earlier, we shared how Hida Sangyo implemented One-Piece Flow for one of its furniture production lines in order to eliminate the waste of Stagnation. Now we will explore how they successfully boosted their productivity in their table-making process by intentionally bringing their processes closer together in order to completely eliminate the acts of both retrieving and placing materials.

Before Kaizen, two of their processes (Process 9: circumference machining and hole drilling, and Process 10: polishing) were located very far apart from one another, which required workers to transport items on carts between them. Since machines performed Process 9 automatically, workers only had to set the job into the machines and push the start button. After activating the machines, workers would immediately start performing other manual labor tasks while the machine worked. However, workers would frequently experience idle time, as they finished performing the tasks that could only be accomplished manually well before the machines completed their jobs. Then, after the machining was complete, the workers had to transport the processed jobs to Process 10. In addition, each work-in-process tabletop had to be loaded onto a cart after it was worked on, then transported to the next workstation.

Through implementing Kaizen, they decided to change their factory layout dramatically to solve this issue. To begin with, they moved two machines in Process 9 just far enough apart that they could fit one machine from Process 10 between them. Now, when a part was transported via cart to the Fridge area for Process 9, each job was fed into both Process 9 and Process 10 by One-Piece Flow, eliminating one full step. As a result, the number of motions each worker took in putting the part onto the cart and retrieving the part from the transportation cart was reduced from nine down to five times. Furthermore, the people in charge of running Process 9 were able to now help workers in Process 10, instead of standing idly waiting for Process 9 to finish (Figure 9.13).

The result was that this particular table line became able to complete 48 tables per 8-hour day, where before it took them 10 hours to complete all 48, in which 2 hours was overtime. This meant that their productivity had risen by 120%. Hida Sangyo has continued to apply Kaizen to their production floor in order to keep up with the true customer demand. As of the writing of this book, they have increased their production capacity to 56 tables in an 8-hour shift, with no overtime required.

The Waste of Transportation in the Shipping Department—Marugo Rubber Industries

Mr. Kenji Takeda of the management division of Marugo Rubber Industries, a company that processes automobile components, was having difficulties with his workers having to repeatedly reload items for transportation. These items were pulled from the Store area for the processing department.

Before they began implementing Kaizen, there were three shipping destinations (A, B, and C). The appropriate finished products were prepicked by workers according to their destination, then placed in a loading zone. However, workers had a tough time seeing which destination a certain item was supposed to go to, and they had to pick through and read the small print on each shipping label closely. This was a challenge because all the items that were transported to the loading zone then had to be sorted and picked all over again when it came time to load items onto each respective shipping truck.

To solve this issue, Mr. Takeda decided to change how the processed items were placed in a Store area. He changed the process so that items

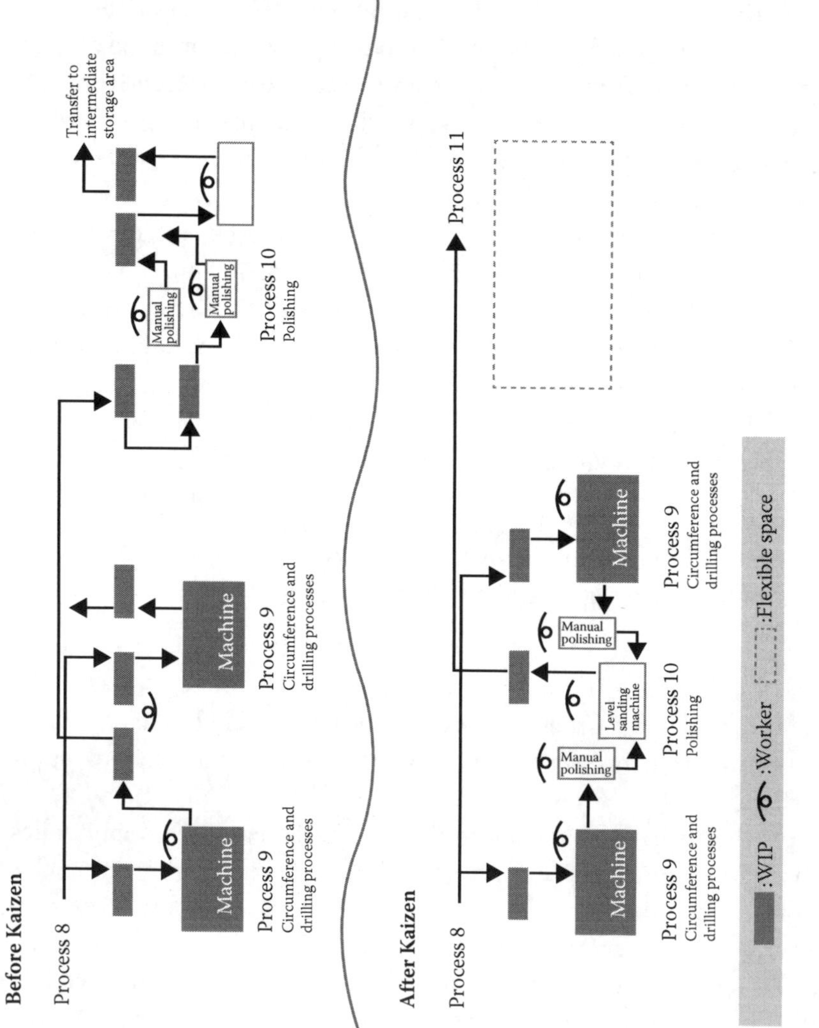

FIGURE 9.13
Removing Transportation waste at Hida Sangyo.

were now placed according to their shipping destinations. From the standpoint of the shipping department, they viewed this area as their Fridge.

After Kaizen, the finished items were transported directly to a Store that was clearly marked for one particular destination. When such items needed to be loaded onto a delivery truck, workers no longer had to search for the right boxes. Instead, they only had to go to a specific Store location to obtain the proper items, which had already been placed according to their shipping destination. Through this Kaizen implementation, the factory freed up 37 square meters of space, reducing the work area by 35%. Time spent sorting and picking was reduced to 45 seconds per case, down from 2 minutes per case before Kaizen (a 62.5% time reduction) (Figure 9.14).

Eliminating the Waste of Searching—Kyowa Manufacturing Co.

Kyowa Manufacturing Co. needed to eliminate the need for searching for parts transported from polishing to cleaning, as this was wasting many hours for their employees each week. Before Kaizen, polished parts from both their internal polishing process line and the external polishing processes from their contracted suppliers were placed randomly in the storage area. This meant that workers had to spend a lot of time looking for the right parts to be fed into the cleaning process. The required parts could only be retrieved after moving away other parts, which were blocking the way, and once the appropriate parts were found, everything else had to be put back where it originally was.

In order to solve this challenge, Mr. Kazushi Kouge, assistant manager, implemented the "Store" concept. Workers would only place an item where it was made or processed. In addition to the creation of a Store, the workers also created two separate areas, one for their internal polishing components and one for their external polished parts, which was further broken down into the three lines they came from (A, B, and C). They also created access aisles around each Store location so that other items did not have to be moved or rearranged for workers to access the necessary items. This allowed workers to easily understand which polishing line was processing the most number of items that needed to be fed into the washing process.

As a result of these improvements, the time required for retrieving the right item was reduced from 1 minute to less than 30 seconds. As 96 products went through the polishing process, the total time saved was 48 minutes per day, which equates to roughly a 10% reduction in time spent searching for items.

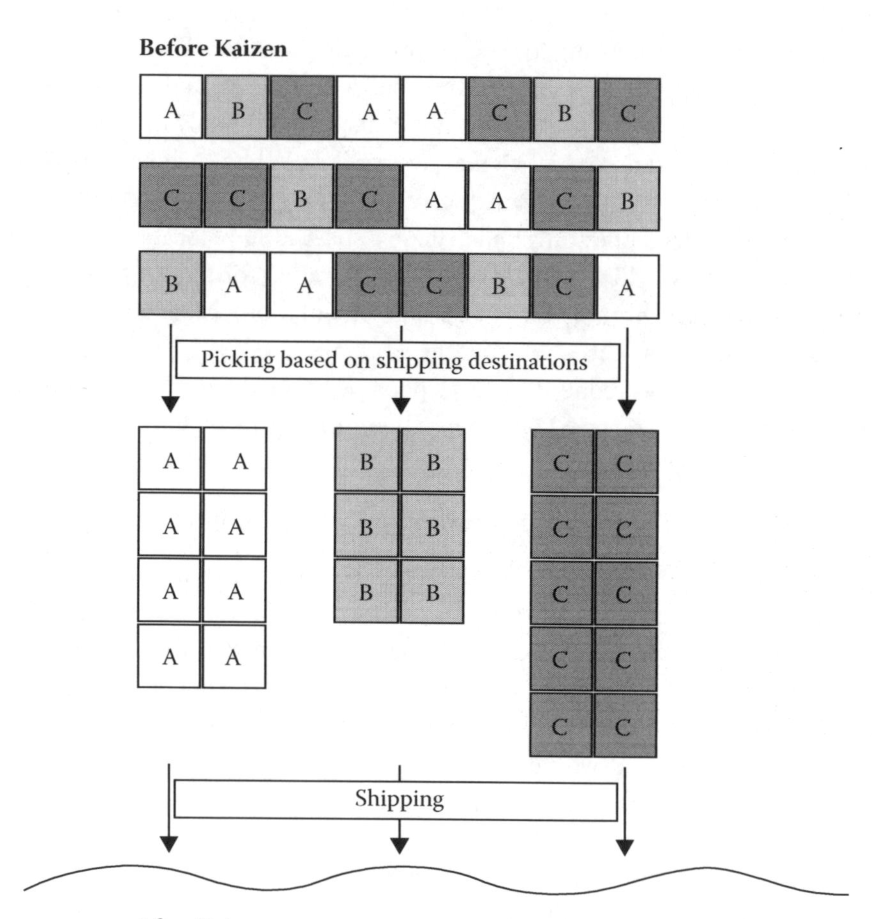

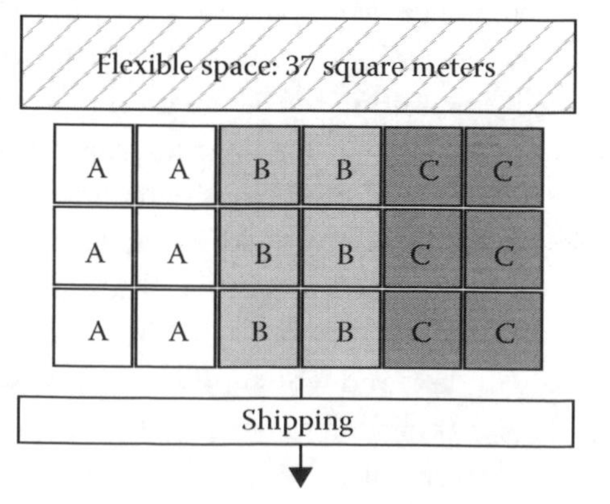

FIGURE 9.14
Before and after Kaizen at Marugo Rubber Industries.

Creating Flexible Workers and Eliminating Rework

A company that manufactures automotive components implemented Kaizen with a focus on eliminating the waste of Transportation in their shipping process. Their process flow before Kaizen was as follows (Figure 9.15):

- Load finished products onto a transportation cart from the Store area designated for storing the finished products
- Move the transportation cart to a shipping lane
- Reload finished products onto the pallets in a shipping lane, emptying the transportation cart

Through Kaizen, they have altered the pallets in their shipping lanes and have attached casters to each pallet. Now, the workers are able to load finished products directly onto the wheeled pallets from the finished product Store as follows (Figure 9.16):

- Load finished products onto the shipping pallets
- Wheel shipping pallets into the appropriate shipping lanes

As a result, the time spent working with each pallet has been reduced from 4 minutes to 1.5 minutes, a 62% reduction in time. As well, the number of employees per shift has been reduced from three to two, creating

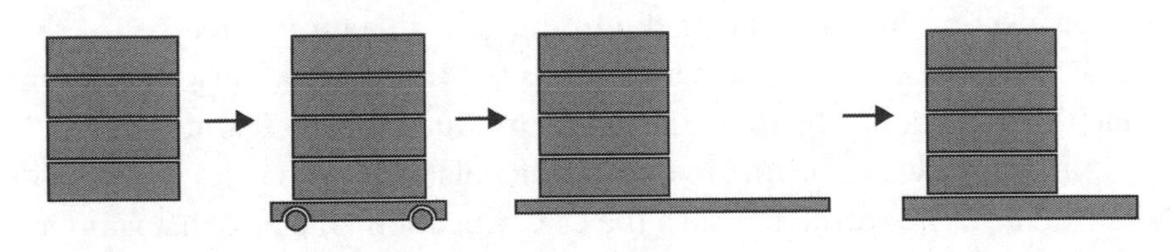

FIGURE 9.15
Shipping process transportation before Kaizen.

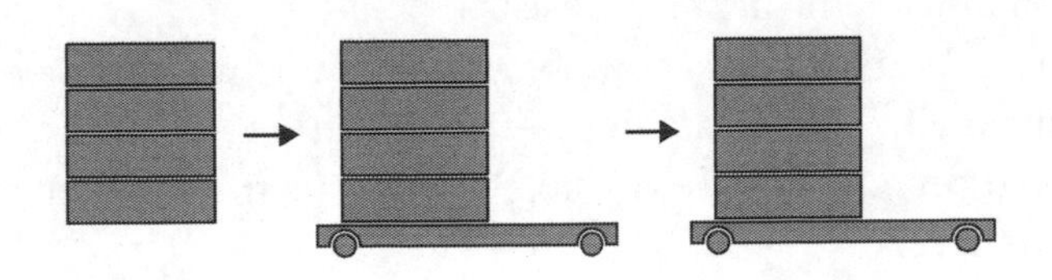

FIGURE 9.16
Shipping process transportation after Kaizen.

Before Kaizen:	1 pallet per 4 minutes
After Kaizen:	1 pallet per 1.5 minutes
= Savings of 2.5 minutes per pallet (62.5% reduction)	

Before Kaizen:	3 workers over 2 shifts
After Kaizen:	2 workers over 2 shifts
= Creation of 1 flexible worker per shift (2 in total)	

FIGURE 9.17
Total savings in shipping process transportation after Kaizen.

one new flexible worker. Since the factory operates in two shifts, the numbers achieved are, in actuality, doubled (Figure 9.17).

Reducing the Distance of Transportation—Tirol-Choco

Tirol-Choco Company was experiencing a difficult time having to transport melted chocolate (a raw material) to the molding process each day. The production line leader, Mr. Yasunori Murakami, noticed that the workers who were in charge of transporting the material, and the workers who were operating the crystallization machine, had some idle time between jobs.

Before Kaizen, the melted chocolate was transported to the back-side of the production line and then fed into the crystallization machine. At the same time, the chocolate was also deposited into another crystallization machine, located in the front-side of the production line (Figure 9.18).

To fix this issue of having to deposit chocolate on both sides of the production line, the workers turned the crystallization machine that was facing toward the back of the line around so that both machines now faced the front of the production line. Transportation pipes were installed on the machine to convey the crystallized raw material to the back of the production line for the next processes.

As a result, Worker B was now able to take over the entire job of transporting the melted chocolate to the crystallization machines, freeing up Worker A to become flexible and able to move to another area of production (Figure 9.19).

Mr. Murukami has continued to work with the employees at Tirol-Choco to eliminate wastes and over the last 2 years has reduced the number of

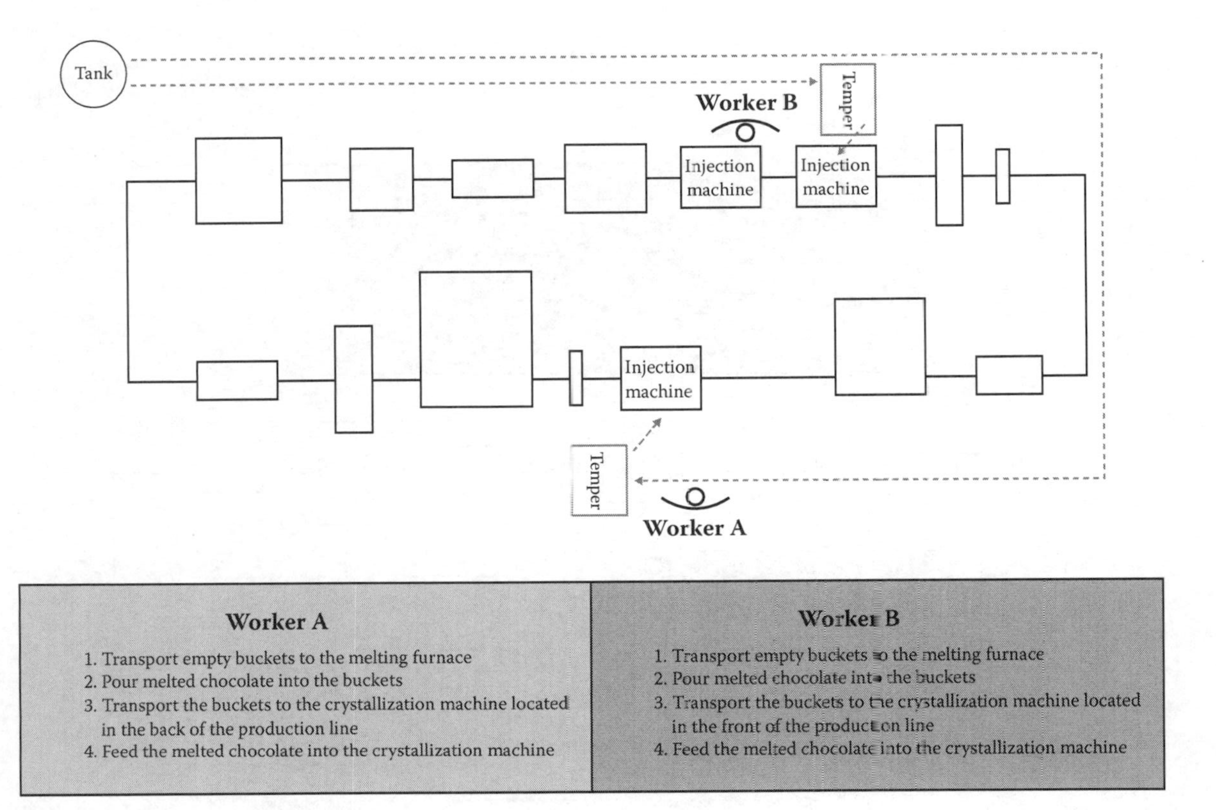

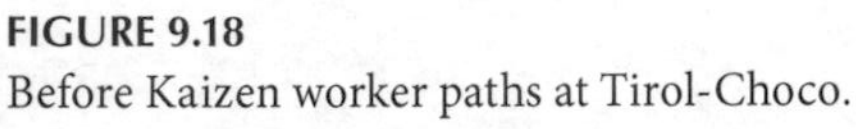

FIGURE 9.18
Before Kaizen worker paths at Tirol-Choco.

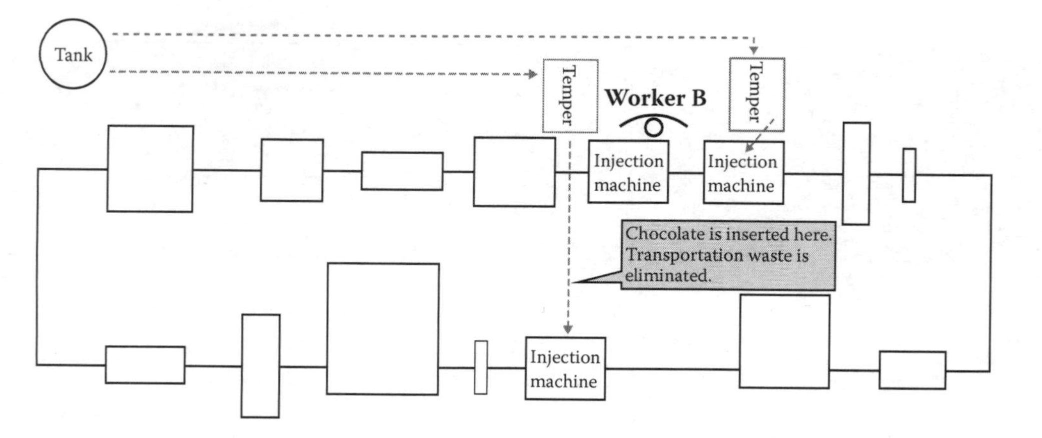

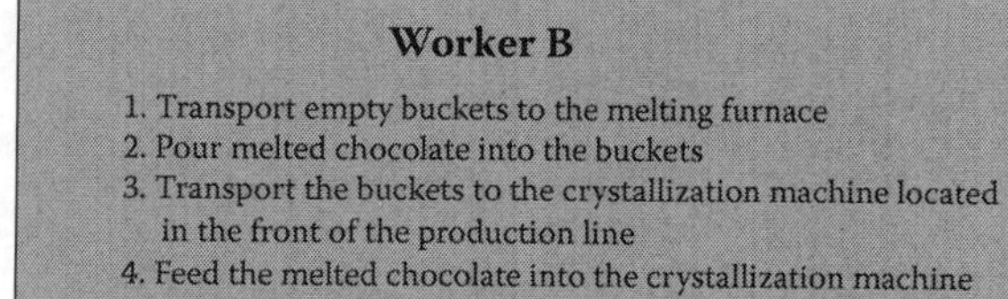

FIGURE 9.19
After Kaizen worker paths at Tirol-Choco.

necessary workers in that process from 15 down to 8, creating 7 flexible, multiskilled workers.

TURN YOUR TRANSPORTATION INTO A MILK RUN

When you need to collect items from many different locations and bring them to a single fixed location, exercising a cyclic transportation method, such as a Milk Run Delivery System, is extremely effective in reducing the waste of Transportation. A Milk Run Delivery System came from how the dairyman delivered bottles of milk to each household every morning in an efficient manner, without any unnecessary transportation.

As shown in Figure 9.20, transporting the finished products from each production line to the shipping area is necessary. In this scenario, the load is usually full on the way to the shipping area, but empty when it goes back to the respective production line. As a result, half the total transportation time spent does not involve any moving of items—a waste!

Implementing a Milk Run transportation system to promote cyclical transportation allows workers to transport items to the shipping area, after having picked up only the necessary items from the Store located beside each production line. This method can reduce the time that is spent moving without a load (Figures 9.21–9.23).

In order to eliminate any movement of the transportation costs without a load, workers can transport the necessary parts to each respective Store area for production lines on their way back from the shipping area.

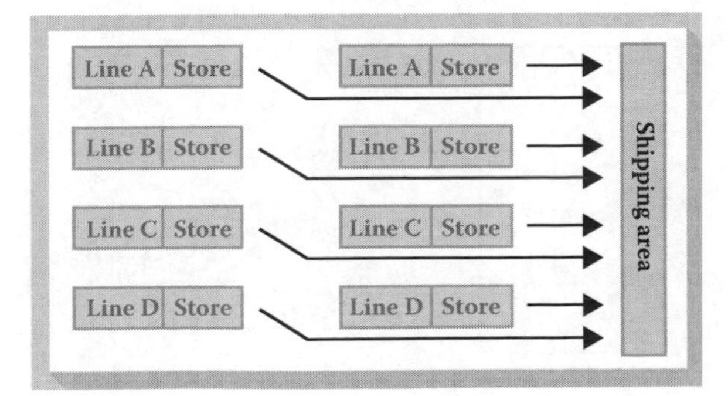

FIGURE 9.20
Milk Run 1.

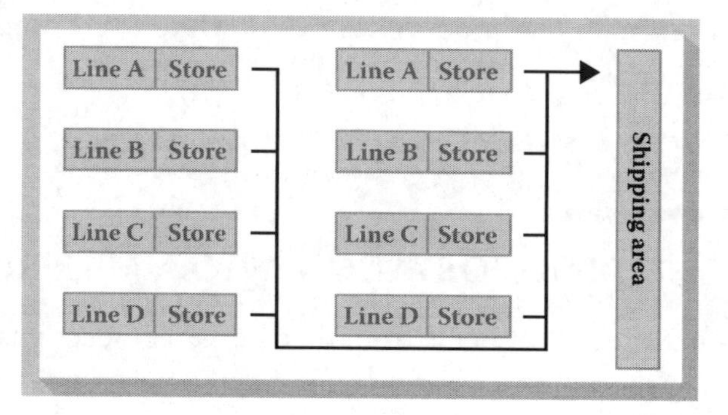

FIGURE 9.21
Milk Run 2.

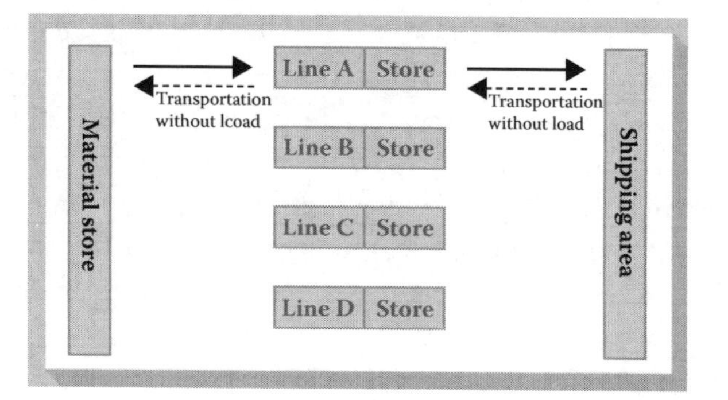

FIGURE 9.22
Milk Run 3.

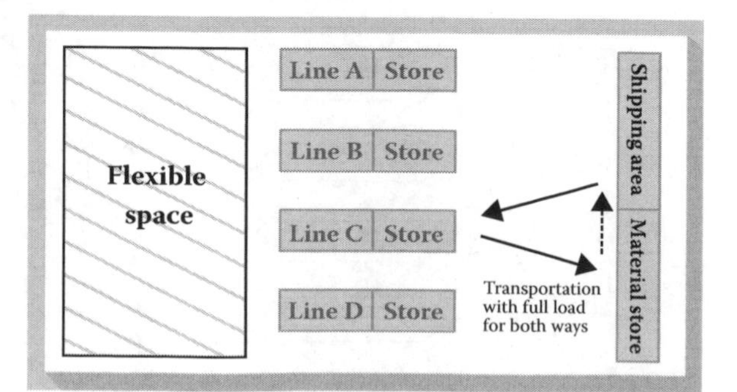

FIGURE 9.23
Milk Run 4.

Implementing a Milk Run System—Unitaka's Okazaki Plant

Unitaka Ltd. is one of the largest textile producers in Japan, and its main plant, the Okazaki Plant, has an operations area of more than 300,000 square meters. This requires an extensive amount of time spent traveling from one place to another within the facility. There are 17 different divisions within the facility, and the person in charge of each division would deliver outgoing mail and pick up incoming mail from a centralized mail distribution center. Each person typically visited the mail center twice a day, and each trip took at least 20 minutes to complete. This meant that each person wasted 40 minutes a day, and if you take a step back and look at the facility as a whole, this adds up to 680 minutes of time wasted on transportation, equal to 11 hours 20 minutes.

One of the section heads, Mr. Masaru Kondo, decided to implement a cyclic mail pickup system and designated a person from the administrative division to make two mail pickup trips per day across the facility. In short, they deputized their own internal mailman. One pickup service for the entire facility took only 1 hour, and overall only 2 hours per day were spent delivering and picking up mail. This completely eliminated the need for representatives from each division to travel to the mail center, and it saved a total of 9 hours 20 minutes each day as a result.

CONCLUSION

By taking some of the lessons from the companies we outlined in this chapter, we hope you can begin to see new ways of eliminating waste in your facility. Once you open your eyes to the true amounts of wasted time, energy, and motion, you will discover a world of improvements to be made.

10

The Kaizen Pyramid

Now that you have been exposed to different ways of seeing and eliminating waste, and you have put some thought into your management mindset and how you can articulate and illustrate your company values, you are ready to begin considering the meaning of True Kaizen.

At the start of this book, we introduced the idea of Kaizen. Now that you have seen many examples and have a better understanding of that idea, we are ready to begin expanding your understanding on the idea of Kaizen. Where does a Kaizen culture come from, and how can you utilize tools to motivate your organization to even greater levels of success?

TOP–DOWN OR BOTTOM–UP?

There are two main approaches to eliminating waste in an organization. One approach is the top–down approach, which comes from the upper leadership/management team, based on achieving an ultimate organizational goal or objective. The other is the bottom–up approach, with full participation from the employees in an organization. This approach encourages an execution of each small idea to get closer and closer to realizing an organizational goal.

Business owners and executive managers whose roles are to formulate effective strategies through visualizing an ideal state and then deploying policies to achieve that state often carry out the top–down approach. One distinct characteristic of a top–down approach is that it can yield a tremendous amount of results at once. However, the weakness of this approach is that most managers tend to make important decisions by examining documented data, such as key performance indicators, instead of visiting a shop floor to fully understand the true cause of problems. Therefore, when managers attempt to implement their best ideas as planned, they are often confused by the gap between their understanding of a problem and the reality of a situation. The difference is often so large that no anticipated results can be obtained, and their strategies need to be altered significantly in the end.

On the other hand, a bottom–up approach, which arises from a shop floor, can surely and gradually give more power to the employees by implementing ideas that accurately reflect the reality of daily work, and are based on practical and empirical information. Their solutions often do not need to be revised during implementation because the workers already know the effect it will have.

However, solutions from a bottom–up approach usually achieve smaller results, though there are more of them than compared to a top–down approach, because small ideas from the shop floor may not directly relate to their manager's ideal state based on corporate strategies (Figures 10.1 and 10.2).

SO WHICH APPROACH IS BETTER?

The goal of true Kaizen is to cycle, essentially connecting top–down strategy with bottom–up daily actions. Kaizen is the mechanism to connect everything, like gears in a watch. Kaizen provides the ability for the organization to close the strategic gap, while showing every employee how his

Top–down vs bottom–up approach							
Planning and Execution		**Perspective**		**Result of Kaizen**		**Continuity**	
Top–down approach	Bottom–up approach	Top–down approach	Bottom–up approach	Top–down approach	Bottom–up approach	Top–down approach	Bottom–up approach
Business owners, managers	Shopfloor leaders, workers	Management strategy, ideal state	Daily operation, task at hand	Extensive	Small	Major modifications and redirection are required	Easy to sustain

FIGURE 10.1
Top–down vs. bottom–up focus.

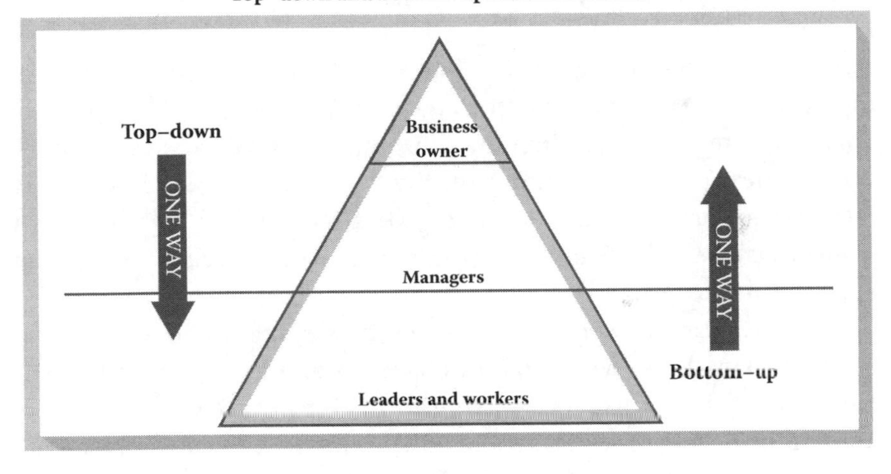

FIGURE 10.2
Top–down and bottom–up one-way communications.

or her contribution allows the organization to meet the challenges of the economy and business environment. We call this The Kaizen Cycle.

The first step in changing the shop floor is a top–down method. When PEC coaches a Muda-tori (Kaizen Waste Elimination) Workshop on the shop floor, their Kaizen coaches take business owners and top managers with them to visit different areas of the shop floor. When the shipping department is reached, they are asked if the final products are being manufactured in the exact quantity the customers have purchased or if there are issues of overproduction and/or underproduction.

By looking at these situations, the business owners and top managers will be able to quantify the amount of Stagnation waste that is associated with finished goods inventories and the waste of Transportation for moving such items to the shipping area. This will also expose them to the real life situations in the business. Managers need to see and connect with front line issues in order to understand strategic intent.

Once these have been addressed, they are then taken back through the entire production process, backward from shipping through all upstream processes, based on the shipping requirement for a particular day. This makes it easier to see and identify wastes in the business.

Cycle Time is calculated based on a shipping volume and available work hours, and the business owners and top managers begin immediately eliminating wastes as they are identified. In this way, PEC trainers use the managers to begin encouraging changes and achieving positive results.

Involving upper management in Kaizen this way triggers the interest of the shop floor leaders and, most importantly, the shop floor employees. Soon they are all brought into the Kaizen process with the upper management teams. After repeating the Kaizen activities, employees start identifying wastes on their own and proactively eliminating them. This transforms the culture of the shop floor and establishes a bottom–up approach to Kaizen.

Implementation of Muda-tori in a continuous manner in the business at all levels will increase overall productivity and leads to a better synchronization of processes. As this transformation occurs, business owners and top managers are able to witness these important changes by taking themselves down to the shop floor. This will also give them more confidence in the capacity of each employee. The stronger a shop floor is with a genuine bottom–up culture, the more it raises the level of organizational operations and becomes aligned with the top–down management in order to truly achieve an organization's strategic goals.

Clear goals are deployed from a top management level and are cascaded down to all the levels throughout the organization until they reach the shop floor. However, the methods for achieving those goals are left in the hands of the employees. This is how you have both a top–down and bottom–up approach. In this way, each department and employee diligently works toward realizing the final goals through daily collaborations (Figure 10.3).

Creating a Kaizen culture and deploying strategies via only a top–down management approach cannot achieve desirable results, as the workplace does not yet have enough capacity to accommodate new requirements, no matter how great and comprehensive their planning may be.

As in sports, an individual's performance level must be trained in parallel with fostering teamwork ability. You might have the world's greatest quarterback, but if he does not have any high-level receivers to throw to, he will never complete a pass.

Similarly, in business, an individual's performance level represents the strength of the shop floor. When it remains the weakest point of your organization, it is extremely hard to achieve desirable goals due to the inability to execute plans successfully. In this case, it is extremely important to bring your focus back onto strengthening the shop floor through training, so that it can acquire the capacity to effectively put your ideas into practice.

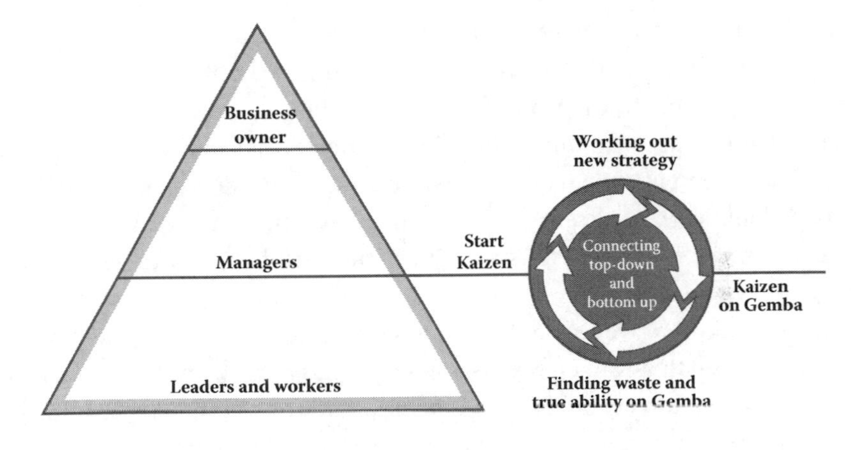

FIGURE 10.3
True Kaizen: The Kaizen cycle.

Even though a bottom-up approach can maintain a significant level of strength in the business, people often tend to focus excessively upon what is happening within their department or section of the business, and they may become indifferent to areas outside of their responsibilities.

Employees in this scenario may implement ideas that are not aligned with the company's strategic goals, as they do not understand where top management is coming from with the goals. Therefore, no matter how hard workers attempt to execute their great ideas, their Kaizen efforts may not be in the best financial interests of the company.

To create a strong marriage between a top–down and bottom–up management, top managers must educate and train all employees in the business by being down on the shop floor with them and understanding the situation from experience, not numbers on a page. Through this approach, workers will have opportunities to interact with top managers and begin to see top management's perspective, allowing them to grasp the direction of where the company is headed. Workers will then adjust their own Kaizen direction and begin applying Kaizen based on this correct alignment with company goals. In doing so, they become part of the Kaizen Pyramid.

RESULTS AND THE KAIZEN PYRAMID

Allow us to explain the truth behind Kaizen: that Kaizen is for everyone at all layers of the business, from the shop floor to the office. It builds a unified perspective that is grounded in action, while at the same time connecting strategy. Too often, managers tend to shoot for achieving big results in the shortest period of time and only ask for Kaizen ideas with the potential for bringing in quick and substantial benefits. Therefore, their management style will end up becoming a top–down approach.

Let's look at the concept of the "Kaizen Pyramid" to see how we can break this cycle of big results in a short period of time, and get to total engagement by all.

Conceptually, as we look at the Kaizen Pyramid (Figure 10.4), starting at the top, we see that Kaizen results are fewer, and actually not as impactful, as they happen too far in between implementations. As well, these Kaizens are not sustainable because, after all, there are only so many times you can completely overhaul your operations.

On the other hand, as we move down to the bottom of the pyramid, Kaizen results become smaller in nature but more impactful because of their independent, frequent nature. These Kaizens create change without escalation or stagnation while waiting for managers to make decisions. This allows the organization to make changes at all levels and react to real demands, both internally and externally, based on the organization's strategy and what that strategy means to each respective layer of the organization.

Therefore, as long as you are only thinking about big Kaizen results, your Kaizen Pyramid will remain small and fleeting.

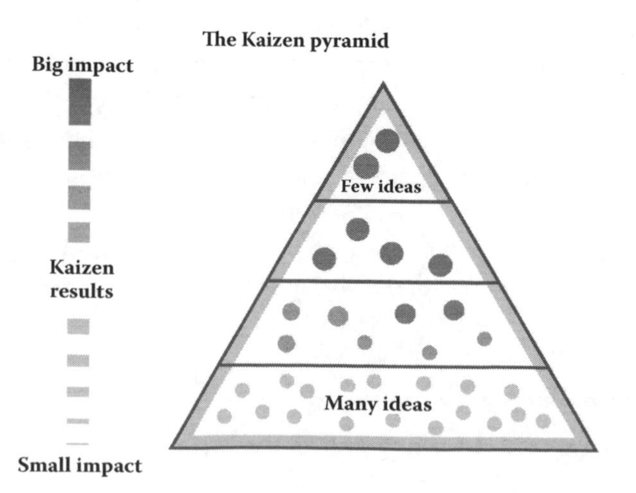

FIGURE 10.4
The Kaizen Pyramid.

THE AIDMA LAW

To further understand this concept, imagine you are a car salesman. What would be the best way to sell cars? Of course, the best way would be to find people who already want to buy a car. The problem here is how you go about finding these people. There is a great model for consumption behavior by Roland Hall called the AIDMA law. AIDMA stands for "attention," "interest," "desire," "memory," and "action" (Figure 10.5).

Taking the scenario of selling cars, from the seller's perspective, the first step is the cognition stage, which is when you gain the customer's attention through advertising. After generating interest and awareness, the next step is the affect stage, which is meant to influence the buyer's emotion by inviting the buyer to a showroom for a test drive. This is meant to generate a strong desire in the buyer to own the vehicle and instill the true value of the car in their mind, which sparks a memory of that "new car" feeling.

The final stage is the action stage, which is to have the consumer actually purchase and use the product. This is how you effectively sell a car. There are more consumers who have a desire and are determined to own a car than customers who are just looking to go through the action of purchasing. Similarly, more customers are interested in cars than customers who desire a car.

As you begin to describe or draw out these types of connections, you will end up with a pyramid shape. Instead of simply collecting people who want to purchase a car in the shortest amount of time, focus on reaching to customers who are simply interested in what cars can do to better their lives so that, as you climb up the pyramid one step at a time, those people will have a strong desire to buy cars from you (Figure 10.6).

The AIDMA law	
Stage	**The buyer's perspective**
Cognition stage	**A** : *Attention*
	I : *Interest*
Affect stage	**D** : *Desire*
	M : *Memory*
Action stage	**A** : *Action*

FIGURE 10.5
The AIDMA law.

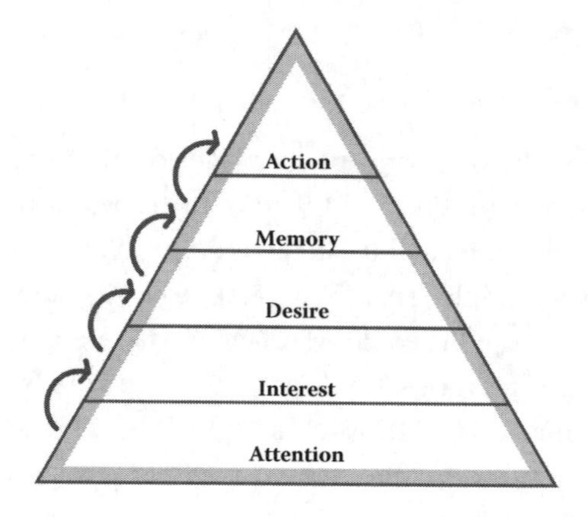

FIGURE 10.6
Climbing up the pyramid of the AIDMA law.

Fundamentally, following the AIDMA rules and making sincere efforts to accumulate customers who are genuinely interested, you will increase customers who will make an action to purchase.

BUILDING YOUR KAIZEN PYRAMID

To begin building your Kaizen Pyramid, you must first get all the employees who work on the frontline involved in identifying one second of waste. You have to build up their sensitivity to waste and encourage behavior that is not complacent. This will establish a strong sense of teamwork and inspire them to continue practicing Kaizen together. As their Kaizen implementations start to show real results in your facility, the integration of both top–down and bottom–up management will become much stronger and supportive of each other (Figure 10.7).

It is very important to reward, recognize, and celebrate Kaizen, especially at this stage, because people are putting their discretionary effort into every single Kaizen, and this needs to be celebrated.

Follow the cycle of Kaizen processes repeatedly so that your Kaizen Pyramid continues to grow. Together, you will be able to collectively achieve the final goals you have always envisioned for your organization. Build a larger Kaizen Pyramid by expanding its foundation, and the rest will follow (Figure 10.8).

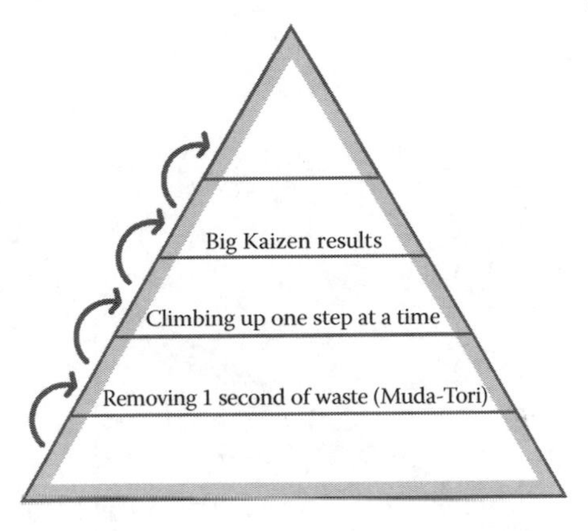

FIGURE 10.7
Climbing the Kaizen Pyramid.

YOUR KAIZEN COMMUNITY

So how do you go about convincing your employees and coworkers to implement Kaizen with you? We often find that people think Kaizen is difficult to understand, at first. So, to help people understand how easy it can be and understand the right mental approach to implementing Kaizen, PEC explains it with a single phrase: "Muda-tori." The purpose of Kaizen is to fundamentally make everything better. Who cannot agree with that?

Promoting this basic idea of Kaizen and Muda-tori with the full participation of everyone involved in the business will make your Kaizen Pyramid bigger and allow you to achieve far greater success.

Throughout this book, we have divided waste up into all sorts of categories: Motion, Transportation, Stagnation, the Seven Types of Waste by Taiichi Ohno—we have defined it and redefined it, but the key is to simply make it memorable.

If a concept is easy for anyone to grasp, this means that it is easier to implement in the business and helps workers identify wastes on the spot. After identifying waste, all one has to do is eliminate it immediately.

For this very purpose, Mr. Yamada has analyzed the Seven Wastes that were formulated by Taiichi Ohno and has simplified them into two major categories of wastes so that anyone in any business can easily understand the concept: Stagnation and Motion/Transportation.

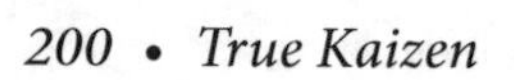

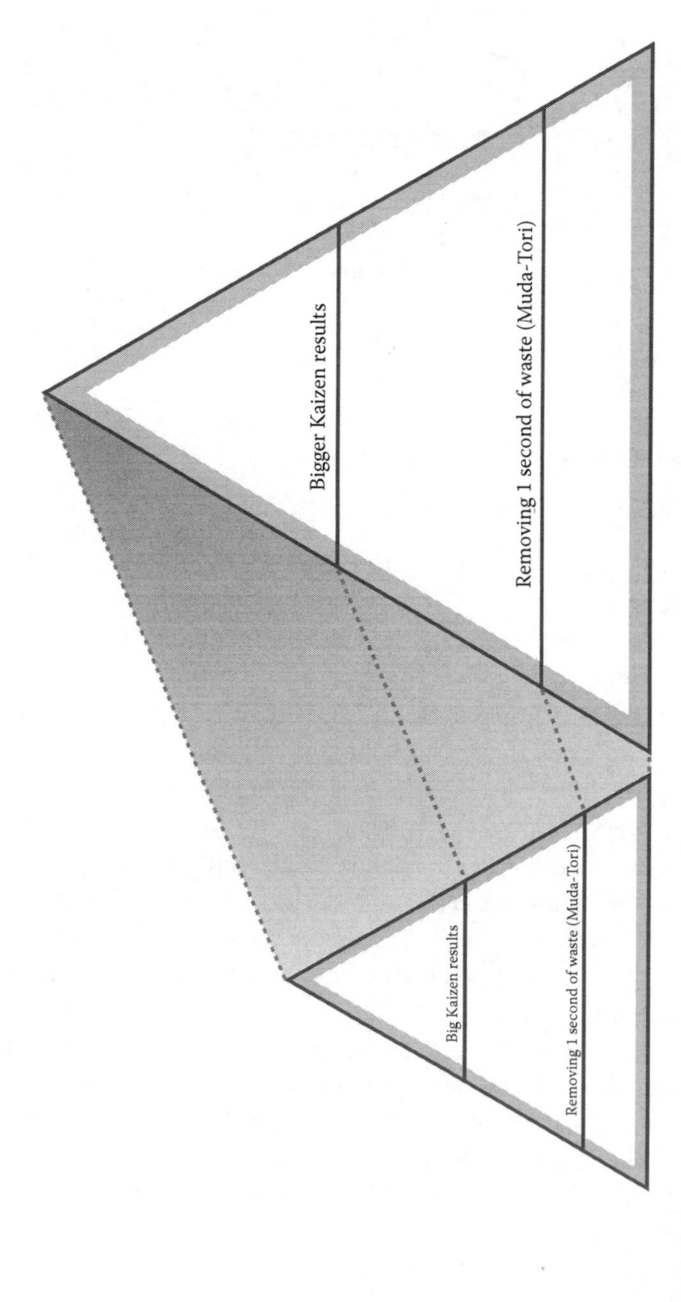

FIGURE 10.8
Expanding the Kaizen Pyramid.

In order to develop people who can identify and eliminate waste successfully, PEC has put an emphasis on developing Kaizen trainers through a Train-the-Trainer course. Trainers who graduate from these courses return to their respective companies and educate their colleagues as the Kaizen leader. Through these internal trainings, more workers become exposed to the joy of Muda-tori and truly begin to implement Muda-tori together. This successful model creates more workers with the same level of passion toward Kaizen and create a true bottom–up Kaizen culture and is something that you too can replicate.

Trainees of PEC's Train-the-Trainer courses attend a 2-day intensive workshop once a month over a 6-month period of time. Trainees receive actual hands on training, at businesses in the group, for four out of their six total sessions. During these training sessions, they are required to not only suggest Kaizen ideas but also to execute their own ideas and achieve positive results, transforming the business by the end of the day. The important focus is not only on acquiring new knowledge but also on experiencing with one's own body what it is like to execute your ideas. This is the most effective way to nurture workers' abilities to identify waste and develop their abilities to take action (Figure 10.9).

In order to become a qualified trainer for Kaizen, they are required to pass four examinations. Examinations include a practical Muda-tori test, a

Overview of PEC's Kaizen Train-the-Trainer course		
Session	**Practical Training**	**Lectures and Discussions**
1	Morale training/ Toyota plant visits	Leadership and Toyota Production System
2	Shop floor training part 1	Identification of Muda and elimination of Muda
3	Shop floor training part 2	Just-in-time concepts and realization
4	Shop floor training part 3	Automation with human intelligence and standard work
5	Shop floor training part 4 and Muda-Tori practical exam	Writing exams, instruction exams, and presentation of Kaizen results
6	Graduation ceremony	Share commitment for the future

FIGURE 10.9
Overview of PEC's Train-the-Trainer course.

written test, an instruction examination, and Kaizen idea execution in a business where each trainee is employed.

In a practical Muda-tori test, trainees identify and remove wastes in the host company and are measured by how well they can eliminate waste on the spot. In the written test, trainees are evaluated by their understandings of the Toyota Production System, as well as the concepts of Muda-tori. In the instruction exam, trainees are tested on how well they are capable of teaching these concepts to other trainees. And in the final stage of the training course, trainees are evaluated on how well they are able to achieve a goal that they challenged themselves with at the beginning of the course. Training as many Kaizen leaders as possible will lead to the expansion of a Muda-tori culture throughout your organization.

Once you begin your transformation, Muda-tori workshops should be held on a regular basis inside your company to reinforce everyone's commitment toward Kaizen, as well as being a platform for sharing new Kaizen concepts and results. With the help of a Kaizen Recognition System and a visual management system, which allows workers to easily share Kaizen results with one another, your employees will keep inspiring themselves to aspire to higher goals. This allows the cycle of Kaizen to continue to evolve, and the Kaizen Pyramid will continue to grow in your organization.

INCREASING THE NUMBER OF KAIZEN TRAINERS

To provide you with a practical example, we will return to the manufacturer of semiconductors, NJR Fukuoka. They had increased the number of Kaizen reports per employee per year from 2.6 reports in the initial year to 26.4 reports 7 years later. They did this through having their own workers participate in PEC's Kaizen Train-the-Trainer course, as well as having internal Kaizen trainers educate other employees within the company (Figure 10.10).

Almost all of the company's top managers, including the president and executive production manager, have obtained the qualifications as Kaizen trainers, and they have proven their abilities for identifying and eliminating waste in many areas of the business.

NJR has also promoted the development of multiskilled workers to the point where over 60% of their employees have become multi-skilled. Even though the company suffered severely from the economic downturn in

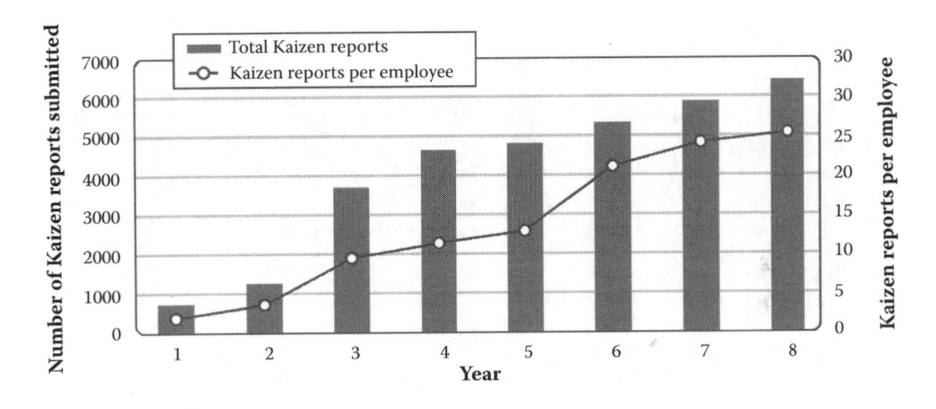

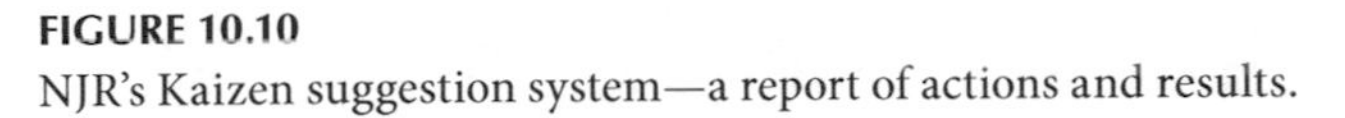

FIGURE 10.10
NJR's Kaizen suggestion system—a report of actions and results.

> "Now, our top managers are able to assist employees in figuring out how they can solve problems through implementing their own Kaizen ideas. As well, they are now able to delegate assistance to other departments that are in need of Kaizen internal training. As a result, our employees have gained more strength in identifying waste, and the number of Kaizen reports completed and submitted by them has drastically increased since we began."
>
> **Mr. Takeshi Honda,**
> *NJR Company Director*

2008 (their sales were reduced by 50%), they were able to sustain profitability through their multiskilled worker program. Their continuous effort behind developing a multiskilled labor force has allowed them to keep producing at a high level of productivity, even with fewer available workers at any given time (Figure 10.11).

TRUE KAIZEN FOR EVERYONE

When you eliminate waste in your business, how much time do you dedicate to it? Compare the time you spend with what PEC's trainees learn in their Train-the-Trainer course.

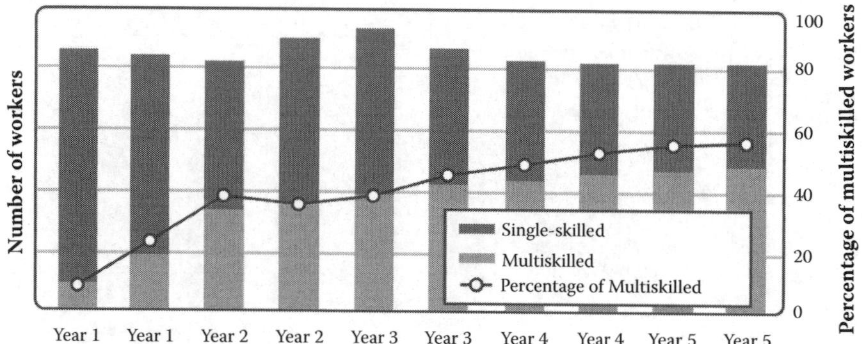

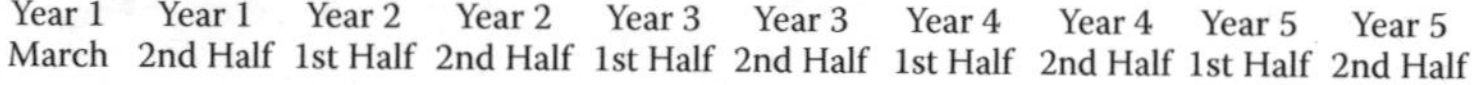

FIGURE 10.11
Multiskilled worker development.

One of the most unique characteristics of PEC's Kaizen Train-the-Trainer courses is that the actual training always takes place where issues are taking place, which is not in a training room. They instruct trainees to not only suggest ideas but also eliminate waste immediately in a speedy and efficient manner. Then, trainees begin motivating themselves to identify waste at the next level and internalize a routine of always acting upon their ideas in order to achieve bigger results repeatedly.

In fact, trainees of this course are given only 3.5 hours to eliminate waste and achieve positive results on the floor. Despite this limitation, they challenge themselves with specific goals such as increasing the productivity of a target area by 20%. After implementation, trainees are required to summarize their Kaizen results in a presentation format and then present these results to other trainees on that same training day.

It is designed to be a very rapid process, where there is not much time to discuss, but it encourages everyone to put their hands to work and show their thinking in terms of flow, building items, changing information, and communicating differently. There is no value in talking about an idea without action, and this time constraint promotes immediate action.

Anyone can suggest good ideas, but to implement these ideas and achieve substantial results, as well as be able to present the outcome in an articulate manner, is extremely challenging. PEC's trainees are more than capable of achieving this requirement after completing a series of Kaizen Leadership Training events.

The idea is that sooner is always better if you want to implement an idea. If you decide to make a proposal for change today, and plan to achieve a result by the end of the month, it would be too late to repair any issues you may be facing at this moment. Instead, you should spend all your energy today and see what can be accomplished within that timeframe. Reduce your Kaizen Cycle (Suggestion, Planning, Implementation) from a monthly basis to a daily basis. Action is much more important than words. Kaizen is not complete without immediate results—if you do not have results, this is not Kaizen.

KAIZEN HAS NO BOUNDARIES

Are the concepts of "Do it First" and "Actions before Words" applicable only to Japan or Japanese companies? No, we believe they are not. In fact, we have had many opportunities to train this unique Kaizen approach to senior managers across the globe. Every year, Collin brings numerous managers onto Japanese businesses' floors to experience True Kaizen. These managers work with Japanese employees and are able to increase productivity by at least 25% in just 1 day. They also receive hands-on Kaizen training directly from Miura, and it is a powerful connection for these managers to understand the true spirit of Kaizen.

Collin also works with Miura, who directs 1-day Kaizen workshops throughout the world and across the United States, to spread this knowledge of True Kaizen.

The first company outside of Japan that Miura began teaching True Kaizen to is Able Engineering, located in Mesa, Arizona. Able Engineering is a great business that has a fantastic organizational culture. We were interested to know if True Kaizen would work outside of Japan, and Able Engineering opened their doors to us. A number of Able's managers had previously been on the benchmarking trips to Japan and were very excited to have Miura visit their business to provide coaching and True Kaizen training.

The 1-day Kaizen Workshop at Able was so successful that they invited Miura back to motivate, encourage, and help them grow their Kaizen culture to the next level to really help them raise the bar. Even in the first workshop, employees were committing to ideas and change, both physically in their efforts and verbally, stating to each other, "We promise you that our Kaizen ideas will increase the productivity more than 120%."

All employees at Able Engineering have become so empowered and confident in their own leadership skills that it has drastically improved the shop floor environment. They are engaged and committed and succeeding (Figure 10.12).

The second company we visited on that first trip is Carl Zeiss Meditec, a manufacturer of medical equipment. Carl Zeiss's facilities in Lebanon, California, already had a Cell Production System in place; however, they wanted to bring down their total assembly line for their process down from 1 hour. Senior managers and the shop floor leaders participated in Miura's Kaizen training and then spent 2 hours together performing Muda-tori and changing the placements of components necessary for their assembly process. As a result, the time spent on Motion and Transportation was reduced by 60% in addition to the total assembly time being reduced by 25%.

At the third company we visited, Rudy's Tortillas, located in Dallas, Texas, we focused on their final packaging process: the area where workers complete boxing multiple packaged products into one finished box every minute. After each box was finished, these boxes were stacked onto other boxes located on a nearby pallet. When we first arrived, employee's travel time followed this process:

- Preparation of a box
- Inserting packaged products into a box
- Transportation of finished boxes on a pallet

FIGURE 10.12
Group photo at Able Engineering.

In order to remove the waste of excess Motion and Transportation, we changed their work layouts dramatically by relocating every necessary item and machine around each worker so that they became able to focus on the task at hand without traveling around the factory looking for items. This work layout change only took us 30 minutes to complete. As a result of this immediate change, the number of steps that workers needed to take to get a job done was reduced by half, which revealed a significant level of uncovered capacity in each worker.

At this point, we advised they make one person a flexible worker and give the other employees more to do by incorporating neighboring production lines, as they became more capable of handling a greater workload without additional resources.

However, the fact that their production lines were located far away from one another made it extremely difficult for people to travel to multiple production lines and return to their home base within 1 minute. While we were facing this challenge, the shop floor leaders proactively suggested bringing the production lines closer together by repositioning the conveyor belts, which eventually merged the two packaging lines into one single line.

As a result of this immediate response to the issue, of the two employees who were previously assigned to each respective packaging line, one was able to do the work of both and so the other became freed up to be a flexible worker—all within only 1.5 hours of work.

In fact, Rudy's Tortillas has been hosting internal seminars and lectures through the help of a Toyota Production System consultant on a regular basis. These "Kaizen Leaders" had an appropriate level of understanding of TPS, but they needed to have the experience of putting their knowledge into practice. Simply put, this company knew of Kaizen but could not practice it until we had the chance to come interact with them and put their knowledge into hands-on learning.

Did either of these American facilities have some secret insight into the Japanese mindset? No. They were filled with workers and managers, just like you, who learned the idea of True Kaizen for themselves. These ideas are not based on culture, but on practice. Take immediate action as soon as you come up with an idea—this is the way of Muda-tori. Workers can wake up to True Kaizen through experiencing this way themselves.

We were originally concerned that this rapid method of executing Kaizen could only be applied to Japanese companies. However, having been able to realize the same results at three companies within the United States without adjusting the approach has proven that it is applicable to any

FIGURE 10.13
Lean Study Mission group photo in Japan.

company, in any industry, regardless of their location in the world. Since this time, we have taken Muda-tori to Canada, Australia, New Zealand, Bahrain, Argentina, and back to the United States again and again, with greater successes being achieved each time. We believe that Kaizen is not cultural but is built into the DNA of everyone (Figure 10.13).

CONCLUSION

Keep promoting True Kaizen through your own actions and become a strong leader on your own behalf. Every human can, and should, thrive and grow. Follow your Kaizen Cycles and expand your Kaizen Pyramid. We strongly believe every one of us can be full of happiness through our work life and become empowered to lead ourselves in the work we are responsible for.

Conclusion

The challenges of management are complex and, at times, can seem so elaborate that it can be hard to locate the path to success. If nothing else, this book should let you know that you are not alone. This book has outlined many examples of real companies confronting the same challenges you face on a daily basis.

Some of these ideas will take time to implement; shifting to an hourly management schedule does not happen overnight, but with persistence, you can make it happen. The key, as always, is the true meaning of Kaizen. When you tap into the creative power of everyone in your organization, there is no challenge that cannot be overcome.

When we set out to write this book, we were on a mission. Looking around at the other books on Lean management, all we could find were technical books about using different practices or a number of techniques. They did not speak to the heart of what it means to create a Kaizen Workplace. They did not address the true meaning of Kaizen, nor did they understand Mr. Ohno's intent and purpose of Kaizen. As a team, we wove this book together, each contributing our ideas for Kaizen—ideas that work—organizational design that is engaging and sustainable, as well as examples from real managers, businesses, our coaching, and our education.

Within every single person, there is the desire, capability, and capacity to make work more fulfilling. By using the principles outlined in this book, you will be able to unlock each person's willingness and desire to be engaged at work. You can make this happen! 100% engagement is achievable no matter what anyone else says. So don't hesitate, seize this moment, educate yourself, do it, and start your journey toward the true meaning of Kaizen.

Miura and Collin

References

Jorge Bucay, *Dejame Que Te Cunte*, Del Nuevo Extremo, 1999

Richard L. Daft, *Management*, Eleventh Edition, South-Western, Cengage Learning, 2014

Walter Davis, *Organizational Behavior: Human Behavior at Work*, 9th Edition, McGraw-Hill, 1993

Eugene Goodson, *How to Read a Plant Fast*, Harvard Business Review, 2002. https://www.hbr.org/2002/05/read-a-plant-fast

Ray Kurzweil, *The Singularity Is Near*, Penguin Books, 2005

Hitoshi Yamada, *Hitome de Wakaru, Sugu ni Ikaseru, Kiso kara Wakaru Kaizen Leader Yosei Koza*, Nikkan Kogyo Shinbunsha, 2008

Hitoshi Yamada, *Forging a Kaizen Culture*, Enna, 2011a

Hitoshi Yamada, *Kaizendamashii no Sakebi*, Nikkan Kogyo Shinbunsha, 2011b

http://en.wikipedia.org/wiki/Frederick_Herzberg

http://en.wikipedia.org/wiki/Hoshin_Kanri

http://en.wikipedia.org/wiki/John_Kotter

http://en.wikipedia.org/wiki/Metcalfe%27s_law

http://en.wikipedia.org/wiki/W._Edwards_Deming

http://www.edelman.com/p/6-a-m/2014-edelman-trust-barometer/

http://www.prisonexp.org/

http://www.theatlantic.com/magazine/archive/1982/03/broken-windows/304465/

http://www.wsj.com/articles/SB10001424127887324178904578340071261396666

https://en.wikipedia.org/wiki/Hawthorne_effect

https://hbr.org/1974/09/skills-of-an-effective-administrator

https://hbr.org/2013/05/what-is-organizational-culture

SFGate.com, http://www.sfgate.com/jobs/salary/article/2013-Wasting-Time-at-Work-Survey-4374026.php, 2014

The Grindstone, http://www.thegrindstone.com/2012/05/31/career-management/you-are-actually-only-productive-for-29-hours-per-work-week-706/

Index

Page numbers followed by f and t indicate figures and tables, respectively.

N

U

V

W